DENSE MOLECULAR GAS AROUND PROTOSTARS AND IN GALACTIC NUCLEI

European Workshop on
Astronomical Molecules 2004

Edited by:

W.A. BAAN and Y. HAGIWARA
ASTRON, Dwingeloo, the Netherlands

H.J. van LANGEVELDE
JIVE, Dwingeloo, the Netherlands

Reprinted from *Astrophysics and Space Science*
Volume 295, Nos. 1–2, 2005

Springer

Library of Congress Cataloging-in-Publication Data is available

ISBN 1-4020-3038-X

2003055495

Published by Springer,
P.O. Box 17, 3300 AA Dordrecht, The Netherlands.

Printed on acid-free paper

Printed in the Netherlands

TABLE OF CONTENTS

STATISTICAL APPROACH

DIAGNOSTICS OF YSO

DIAGNOSTICS OF AGN/STARBURSTS

FOREWORD

The European Workshop on Astronomical Molecules was held at 'The Conference Building' in Zwolle, the Netherlands, on 18–20 February 2004. The idea behind the workshop was to bring together astronomers studying similar processes in different astrophysical environments; masers and dense molecular gas around young stars and galactic nuclei. There is considerable overlap in physical and chemical phenomena between these environments, with scales ranging from circumstellar to central regions of galaxies.

The workshop became a meeting place of observers and theorists as well as young and senior astronomers. The venue of the meeting facilitated continuous stimulated discussions in a relaxed atmosphere throughout the meeting and after sessions in the hotspots of the cosy town of Zwolle.

This issue includes 12 reviews and 27 contributed papers presented in this Workshop, all of them were refereed by invited speakers or the editors. The quality of the papers is outstanding. The projects discussed here are often in an early stage and undoubtedly their progress has benefited greatly from the opportunity to discuss results with the experts in a wide range of areas. Similarly, we feel confident that the interaction of specialists with different backgrounds will initiate many new projects and collaborations.

The format of the meeting was special since all participants were given the opportunity to make an oral presentation and there were no posters. In addition, the schedule allowed sufficient room enough for off-line presentations and discussions. The program switched often between galactic and extra-galactic objects so that all participants could be directly involved in all discussions. The attendance was indeed uniformly high during the meeting. The focus was always on trying to interpret the latest observations of molecules in dense gas environments, whether in the vicinity of nearby young stars or at active nuclei in distant galaxies. The workshop provided an expectant perspective of the future and the new observational and modeling opportunities ahead.

As editors, we acknowledge the enthusiasm of all participants, the help with the review process, and the timely submission of manuscripts.

Many thanks go to the other members of the Scientific Organizing Committee. In particular, we thank Nanuschka Csonka and Marjan Tibbe who handled effectively all the logistics of the workshop and always were prepared to solve problems.

We are grateful to Hiroshi Imai for his initiative in setting up and organizing the workshop. We also acknowledge the RadioNet Consortium for a grant that allowed young participants to join the workshop.

Tot ziens,
Dwingeloo, August 2004.

Yoshiaki Hagiwara
Willem A. Baan
Huib Jan van Langevelde

Astrophysics and Space Science **295:** 1, 2004.

YOUNG STELLAR OBJECTS (YSO) EVOLUTION

OUTFLOW, INFALL, AND ROTATION IN HIGH-MASS STAR FORMING REGIONS

RICCARDO CESARONI
INAF, Osservatorio Astrofisico di Arcetri, Italy

(Received 16 April 2004; accepted 15 June 2004)

Abstract. According to theory, stars more massive than $\sim 8\ M_\odot$ must form while still accreting material from the surrounding parental cloud: at this stage radiation pressure should reverse the infall thus preventing further growth of the stellar mass. After illustrating the two models proposed to solve this problem ("accretion" and "coalescence"), we review the observational evidence pro/contra such models, focusing on the kinematics of the molecular gas where the massive (proto)stars are embedded as the best tool to shed light on the formation mechanism. Special attention is devoted to the phenomena of infall, outflow, and rotation, concluding that the recent detection of rotating disks in massive young stellar objects is the best evidence so far in favour of the accretion model.

Keywords: star formation, jets and outflows, HII regions, early-type stars

1. Introduction: The Formation of Massive Stars

The term "massive" is commonly used to identify early-type stars, i.e. stars with O–B spectral type. Such a definition is to some extent arbitrary and mostly related to the dramatic increase of the ionising luminosity of the star as its mass exceeds a value of a few solar masses. However, from the point of view of the present review, a more precise definition is in order, as we shall discuss the *formation* process of high-mass as opposed to low-mass stars. Therefore, we shall briefly point out the reason for setting the dividing line between these two types of stars at $\sim 8\ M_\odot$, illustrating the dramatic implications this may have on the formation of stars above this mass and the models proposed to explain such a process.

This review will focus on the role of observations in determining which of the theoretical models proposed to explain the formation of massive stars is correct. After illustrating the observational implications of the different models, we shall summarise the present evidence for infall, outflow, and rotation and discuss the relevance of these phenomena to the high-mass star formation process. Special attention will be devoted to the possible existence of disks in high-mass young stellar objects (YSOs) and to the fundamental role these may play in ruling out or confirming the proposed formation scenarios.

Astrophysics and Space Science **295:** 5–17, 2004.

2. The Dividing Line Between High- and Low-Mass Stars

In a widely accepted scenario, the formation of (low-mass) stars proceeds through inside-out collapse of an isothermal core (Shu et al., 1987). In a simplified description of the process, one may imagine that accretion proceeds onto a dense kernel, the "protostar", which radiates the gravitational energy released in the collapse. To this process two timescale are relevant: the accretion timescale of the envelope, $t_{\rm acc} = M_*/\dot{M}_{\rm acc}$, and the contraction timescale of the kernel, the so called Kelvin–Helmotz timescale, $t_{\rm KH} = GM_*/R_*L_*$. Here we have indicated with M_*, R_*, and L_* respectively the mass, radius, and luminosity of the protostar, and with $\dot{M}_{\rm acc}$ the mass accretion rate. For low M_* $t_{\rm acc} < t_{\rm KH}$, for $M_* \simeq 8\ M_\odot$, the two timescales become equal (Palla and Stahler, 1993), and for larger masses $t_{\rm acc}$ exceeds $t_{\rm KH}$. This has the consequence that stars below 8 $M_\odot$ contract very slowly as compared to accretion, so that once their reservoir of material is exhausted they may slowly contract and heat up until they eventually ignite hydrogen burning, thus reaching the zero-age main sequence (ZAMS). On the other hand, if the reservoir (i.e the parental clump) is large enough, the star will keep growing until the critical mass of 8 $M_\odot$ is reached. Beyond this point the star will rapidly collapse, because $t_{\rm KH} < t_{\rm acc}$, thus reaching the ZAMS still deeply embedded in the accreting environment.

At this stage, the radiation pressure and copious UV flux emitted by an O–B star are bound to halt the collapse and disrupt the surrounding environment thus preventing further growth of the stellar mass. The paradoxical conclusion is that massive stars (i.e. stars above 8 $M_\odot$) should never form. Such a paradox clearly demonstrates that formation theories successfully applied to low-mass stars cannot be trivially extended to their high-mass counterparts.

3. High-Mass Star Formation: Possible Solutions

Eventually, what hinders further growth of a massive protostar is the radiation pressure exerted on dust grains in the infalling material. The problem has been clearly illustrated by Yorke (2004): to make accretion possible, the dust absorption coefficient, $\kappa_{\rm dust}$, must be less than a critical value proportional to the ratio M_*/L_*. This result may be obtained by reducing $\kappa_{\rm dust}$ or by overcoming the radiation pressure with the ram pressure of the infalling gas. Non-spherical accretion is also very effective for this purpose, as it is much more difficult for the radiation to blow away a massive disk. On this basis, two models have been proposed to solve the paradox of massive star formation: *(non-spherical) accretion* with sufficiently large accretion rates (Behrend and Maeder, 2001; Yorke and Sonnhalter, 2002; Tan and McKee, 2002); and *coalescence* of stars with masses below the critical value of $\sim$8 $M_\odot$ (Bonnell and Bate, 2002). As pointed out by Yorke (2004), the latter is an extreme case of accretion with low values of $\kappa_{\rm dust}$, as radiation pressure is totally ineffective against merging stars.

In this contribution, we wish to review the evidence pro or contra these two models. It is thus helpful to compare their implications and see which of these is supported by the observations. Clearly, the most distinctive feature between the two models is given by the fact that isolated massive star formation is impossible in the coalescence model, for which the presence of a large number of stars ($\sim 10^8$ pc^{-3}) is mandatory, whereas this is possible in the accretion scenario. However, the latter predicts that high-mass stars form at the centre of massive molecular clumps which give origin also to a large number of lower mass stars, thus making isolated star formation very unlikely in any case. A much more effective way to discriminate between the two models is given by the kinematics of the gas associated with the forming stars. In fact, both models require the presence of infalling gas which, by conservation of angular momentum, is bound to form flattened, rotating structures, i.e. circumstellar disks, which in turn are very likely associated with outflowing gas, so that bipolar molecular outflows are also expected. However, only the accretion model is consistent with the existence of well-defined disk/outflow systems: in the coalescence model, instead, the disks/outflows associated with the low mass stars should be destroyed during the merging process. It is hence clear that the detection of disk/outflow systems in high-mass YSOs would lend strong support to the accretion model.

4. The Kinematics of Massive Star Forming Regions

The main conclusion one may draw from the previous discussion is that the study of infall, outflow, and rotation in high-mass star forming regions is a valuable tool to shed light on the formation process of early-type stars. In the following we shall discuss the observational evidence for these three phenomena.

4.1. INFALL

Assessing the presence of accretion is a difficult task, mostly because it is only on very small regions that the infall velocities may contribute significantly to line broadening: basically, this is due to the fact that the free-fall velocity increases as $R^{-0.5}$, with R distance from the collapse centre. For instance, if the material is infalling onto a 10 $M_\odot$ star, it will reach a velocity of 1 km s^{-1} at a distance of $\sim$0.09 pc from the star: this is only marginally detectable in molecular lines as broad as 5–10 km s^{-1}, commonly measured in massive star forming regions. In most cases direct evidence for infall is obtained from red-shifted (self)absorbed line profiles (see e.g. Keto et al., 1987; Hofner et al., 1999a) although high angular resolution is needed to unambiguously assess their origin: in fact, inverse P Cyg profiles may be easily mimicked by multiple unresolved clumps with different velocities along the line of sight. Infall can be also detected in the ionised component surrounding a newly formed massive star, as it seems demonstrated by recent observations of the H66α recombination line in the ultracompact HII region G10.6–0.4 (Keto, 2002).

Some authors have found indirect evidence for infall demonstrating that massive parsec-scale clumps where massive YSOs are found are likely not virialised and show density profiles indicating instability (Fontani et al., 2002). Other authors have obtained reasonable fits to the SEDs of massive YSOs only adding a contribution from accretion to the bolometric luminosity of the source (Osorio et al., 1999).

Although no iron clad evidence for infall has been presented so far, the estimated infall rates are 10^{-3}–10^{-2} $M_{\odot}\mathrm{yr}^{-1}$, i.e. up to three orders of magnitude greater than those typical of low-mass star forming regions. Accretion rates that large are sufficient to overcome the radiation pressure by any massive star, thus making the accretion model feasible.

4.2. Outflow

Unlike infall, outflow motions are easy to detect, but are of limited use due to their poor collimation which makes it difficult to assess their origin and hence associate it unambiguously with a given massive (proto)star. Indeed most outflow searches in high-mass star forming regions have been performed with observations of CO(1–0) or (2–1) lines at low angular resolution (Shepherd and Churchwell, 1996; Zhang et al., 2001; Beuther et al., 2002a). These have been sufficient to prove that outflows are as common in high-mass YSOs as in low-mass ones. They are characterised by large outflow rates and masses, which seem to present a continuity with respect to the corresponding values in low-mass YSOs: in fact, Churchwell (2000) has shown that the mass loss rate, $\dot{M}$, correlates well with the source luminosity, L, over six orders of magnitude, according to $\dot{M} \propto L^{0.7}$ (but see Beuther et al., 2002a for a different point of view).

Recently, an ever growing number of observations of selected sources with arcsecond resolution have been performed, which have made it possible to resolve the structure of the flows and better assess their properties. The most relevant finding is that in several cases what appeared to be a single, poorly collimated outflow splits into multiple well-collimated flows when observed with better resolution (Beuther et al., 2002b, 2003). In other cases, the orientation of the outflow axis seems to change from the large to the small scale, which is interpreted as precession of the flow (Shepherd et al., 2000). These effects complicate the picture of outflows in massive star forming regions and make it difficult to identify the YSOs powering them. It is hence very difficult to establish the uniqueness of an outflow in association with a given massive YSO, as implied by the accretion model. In this context, water maser studies at high angular resolution may be very helpful to trace the ejected high velocity gas and thus pinpoint the location of the (proto)star (see Torrelles et al., 1997, 2003). On the other hand, the existence of multiple outflows seems to support the coalescence scenario: however, it must be noted that each of these outflows is much more massive than expected on the basis of the coalescence model, which requires stars below 8 $M_{\odot}$.

4.3. DISKS

In view of the problems associated with the use of infall and outflow to discriminate between theoretical models, it is of great interest to investigate the presence of rotating disks in massive YSOs. Not only their existence may be reasonably foreseen on the basis of basic physical principles such as conservation of angular momentum during the collapse, but they have been directly observed in a large number of low-mass YSOs (Simon et al., 2000). What about high-mass YSOs? Many claims of rotating disks in massive star forming regions are reported in the literature. These are summarised in Table I, where the tracers used and the corresponding references are reported. As one may see, there are basically three different types of tracers: maser lines, continuum emission, and thermal lines. Let us discuss these in some better detail.

4.3.1. *Disks Traced by Maser Lines*

Maser lines are the most suitable to obtain a high angular resolution picture of a circumstellar environment, as they arise from regions as small as a few AU with brightness temperatures which may reach 10^{12} K. These characteristics make them ideal targets for VLBI studies and excellent tracers of the velocity field. However, other problems hinder the use of maser emission: the large time variability of some masers (especially H_2O) and the fact that they often arise in shocked regions, make their observation and interpretation quite difficult and often ambiguous. Even in the best studied cases such as that of Orion KL (Greenhill et al., 2004), where a 3-D

TABLE I

List of tracers used to search for disks in high-mass YSOs and corresponding references

Tracer	References
CH_3OH masers	Norris et al. (1998); Phillips et al. (1998); Minier et al. (1998, 2000)
OH masers	Cohen et al. (2004)
SiO masers	Greenhill (2004)
H_2O masers	Fiebig et al. (1996); Torrelles et al. (1996); Shepherd and Kurtz (1999)
NIR, mm, cm continuum	Gibb et al. (2004)
NH_3, $C^{18}O$, CS, $C^{34}S$ CH_3CN	Keto et al. (1988); Cesaroni et al. (1997, 1998, 1999); Shepherd and Kurtz (1999); Shepherd et al. (2000); Sandell et al. (2003); Beltrán et al. (2004); Zhang et al. (1998a,b, 2002)

view of the velocity field has been obtained with proper motion measurements of the SiO maser spots, it remains unclear whether one is observing rotation, outflow, or infall, or possibly a combination of these. Similarly, H_2O masers have also given ambiguous results. For instance, in Cep A HW2 (Torrelles et al., 1996) the H_2O maser spots are distributed in an elongated structure perpendicular to the axis of a thermal jet arising from a luminous YSO. However, a velocity gradient in the spots is seen both in the direction of the axis and perpendicular to it. It is hence unclear whether one is observing a rotation about the jet axis, expansion along that axis, or infall towards the YSO.

CH_3OH masers are even more ambiguous. In many cases, the maser spots turn out to be distributed along lines or arcs with velocity monotonically increasing along these structures. Several authors (see references in Table I) have interpreted these as Keplerian disks. Under this assumption one may obtain a lower limit to the mass of the central (proto)star from the measured velocity gradient. Although for a few objects this method gives reasonable values for the stellar mass (e.g. 6 $M_\odot$ in W48; see Minier et al., 2000) in the large majority of the cases the masses obtained in this way are less than 1 $M_\odot$. Moreover, De Buizer (2003) has searched for jets in the 2.2 μm line towards a sample of CH_3OH maser sources claimed to trace rotating disks. In all cases where a jet could be identified, the direction of this is parallel to the maser structures, and not perpendicular to them as expected if the masers were tracing disks.

A general conclusion concerning masers as tools to search for disks in massive YSOs, is that they can be of limited use without additional information on the source, such as the direction of an associated jet/outflow. At present, while convincing evidence has been reported on the association between masers and outflows (see e.g. Moscadelli et al., 2000) no unambiguous case of masers in a disk has been found.

4.3.2. *Disks Traced by Continuum Emission*

In low-mass star forming regions, thermal line and continuum emission have been successfully used to trace circumstellar or circumbinary disks, the first and perhaps most striking case being that of GG Tau (Guilloteau et al., 2002). Since continuum emission at (sub)mm wavelengths is proportional to the mass and temperature of the dust (and hence of the gas, given the coupling between the two at large densities), this should be even more effective for massive/luminous YSOs, which one might naïvely expect to be associated with more massive and hotter disks. However, in this case typical distances are an order of magnitude larger than for low-mass YSOs, which makes it very difficult to resolve the disk and disentangle the emission from the disk from that of a possible stellar wind. An example of searches based on the mm continuum emission is that of Gibb et al. (2004), who also combined the information from their BIMA observations of a sample of massive YSOs with previous NIR images (Hoare et al., 1996). Although suggestive of the presence of circumstellar disks, their findings confirm that better angular resolution and sensitivity at mm wavelengths are needed to draw a firm conclusion.

4.3.3. *Disks Traced by Thermal Line Emission*

The obvious advantage of observations of thermal lines with respect to continuum is that these supply us with information on the velocity of the gas. This makes it possible to reveal a rotating disk through its velocity gradient even if the angular resolution may not be sufficient to properly resolve its structure. With millimeter interferometers one may cover a large number of lines from different molecular species, which is of great importance to investigate the kinematics of the molecular cloud on different spatial scales. This is especially important when searching for disks. In fact, the detection of a velocity gradient in a core is not by itself guaranteed for the existence of a rotating disk (as seen in Section 4.3.1 for CH_3OH masers), but if such a gradient turns out to be perpendicular to the axis of an outflow, the disk interpretation becomes very likely. Indeed accretion disks and outflows are believed to be tightly associated: with this in mind, detailed analyses of massive YSOs with outflows have been performed through interferometric observations of core *and* outflow molecular tracers. Once a velocity gradient in the core is observed, the information on the outflow direction is sufficient by itself to establish whether the gradient in the core is due to rotation or instead is just tracing the root of the outflow itself: in the latter case, the gradient in the core should be parallel to the outflow axis, whereas rotation requires perpendicularity between the two.

This technique has been applied successfully in several luminous YSOs, leading to the discovery of a handful of rotating disks. These are reported with their properties in Table II, sorted according to the bolometric luminosity. It must be stressed that in most cases it is unclear whether the luminosity (which comes from IRAS measurements) is indeed originating from a single YSO or a whole cluster: nevertheless, higher luminosities indicate significant differences in the properties of the region observed and hence of the disk associated with it, no matter whether one is observing a circumstellar or a "circumcluster" disk. As a matter of fact, looking at Table II one can see that B (proto)stars are associated with less massive and smaller disks whereas the opposite occurs with more luminous YSOs. With this in mind, in the following disks below a few 10 $M_\odot$ will be referred to as "disks", whereas the term "toroids" will be used for the others. The main reason for using a different terminology is related to the ratio between the mass of the disk and that of the star: while this ratio is around unity for B (proto)stars, in more luminous objects it becomes >10, which proves the disk to be unstable. We thus believe that a distinct terminology is in order for such "fat" disks.

In Figures 1 and 2 prototypical examples of a disk and a toroid are shown. The former illustrates the structure of IRAS 20126+4104 from $\sim$0.2 pc to $\sim$200 AU. On all scales a bipolar jet/outflow oriented SE–NW is clearly seen in a variety of tracers such as HCO^+, H_2, 3.6 cm continuum, and H_2O masers. At the geometrical centre of this flow a compact 7 mm continuum source is detected with the VLA

TABLE II

List of disks and toroids in high-mass star forming regions

Name	$L(L_\odot)$	$M(M_\odot)$	R(AU)	$M_*(M_\odot)$	References
G192.16–3.82	3×10^3	15	1000	6–10	Shepherd and Kurtz (1999); Shepherd et al. (2001)
AFGL 5142	4×10^3	4	1800	12	Zhang et al. (2002)
G92.67+3.07	5×10^3	12	14400	4–7.5	Bernard et al. (1999)
IRAS 20126+4104	10^4	4	1600	7	Cesaroni et al. (1997); Zhang et al. (1998b); Cesaroni et al. (1999); Cesaroni et al. (in preparation)
NGC7538 S	10^4	100–400	30000	40	Sandell et al. (2003)
G29.96–0.02	9×10^4	300	14000	<25[a]	Olmi et al. (2003)
G31.41+0.31	3×10^5	490	16000	<40[a]	Cesaroni et al. (1994); Beltrán et al. (2004)
G24.78+0.08[b]	7×10^5	80–250	4000–8000	20–40[a]	Beltrán et al. (2004)

[a]Upper limit assuming that luminosity originates from one ZAMS star.
[b]Three toroids have been found in the G24.78 star forming region.

in its most extended configuration (Hofner, personal communication): most likely this marks the position of a luminous YSO deeply embedded in the molecular core seen in the mm continuum with the Plateau de Bure interferometer by Cesaroni et al. (1997, 1999). These authors have also mapped the core in CH_3CN, a high density tracer, revealing a velocity gradient perpendicular to the jet/outflow axis, which they interpret as a Keplerian disk rotating about the embedded YSOs. Recent observations in the $C^{34}S$(2–1) and (5–4) lines (Cesaroni et al., in preparation) confirm this hypothesis, leading to an estimate of 7 $M_\odot$ for the mass of the star, and in addition provide us with evidence of infall in the core. We conclude that in IRAS 20126+4104 one is dealing with a circumstellar accretion disk in Keplerian rotation about an early-B (proto)star.

The situation in the case of G24.78+0.08 is significantly different from that of IRAS 20126+4104. Figure 2 illustrates a comparison between the bipolar outflows detected by Furuya et al. (2002) with the velocity gradients found by Beltran et al. (2004) in the cores hosting the YSOs powering the flows. This picture is reminiscent of Figure 1, but in G24.78+0.08 the luminosity and mass of the toroids are almost two orders of magnitude greater than in IRAS 20126+4104. Clearly, Keplerian rotation is impossible as the disk mass is much larger than any reasonable stellar mass. Also, Beltran et al. (2004) find that the masses of the toroids are significantly greater than the dynamical masses required for equilibrium and conclude that one is observing unstable, infalling/rotating structures.

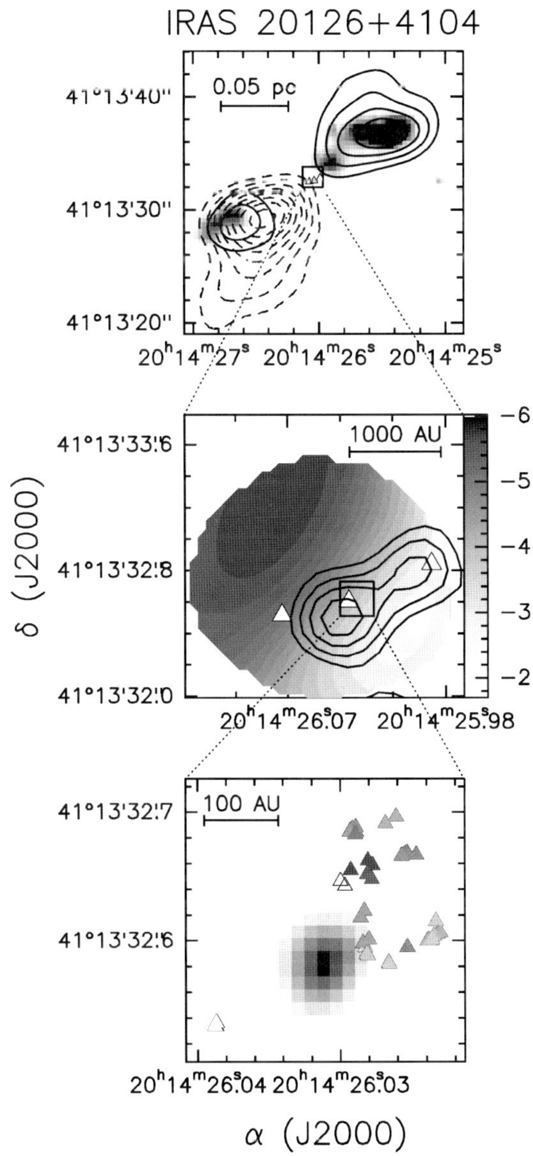

Figure 1. Disk/outflow system in the high-mass (proto)star IRAS 20126+4104. Top: overlay of the H_2 lime emission at 2.2 μm (grey scale) and the outflow map obtained from the HCO^+(1–0) line (Cesaroni et al., 1997). Full and dashed contours indicate respectively blue- and red-shifted emission. The triangles mark the positions of the H_2O maser spots detected by Tofani et al. (1995). Middle: 3.6 cm continuum map (contours; Hofner et al., 1999b) overlayed on a map of the velocity measured in the $C^{34}S$(5–4) line by Cesaroni et al. (in preparation). Bottom: distribution of the H_2O maser spots (Moscadelli et al., 2000; Moscadelli et al., in preparation) compared to a VLA map (image) of the 7 mm continuum emission (Hofner personal communication). The grey scale of the spots ranges from white, for the most red-shifted spots, to black, for the most blue-shifted.

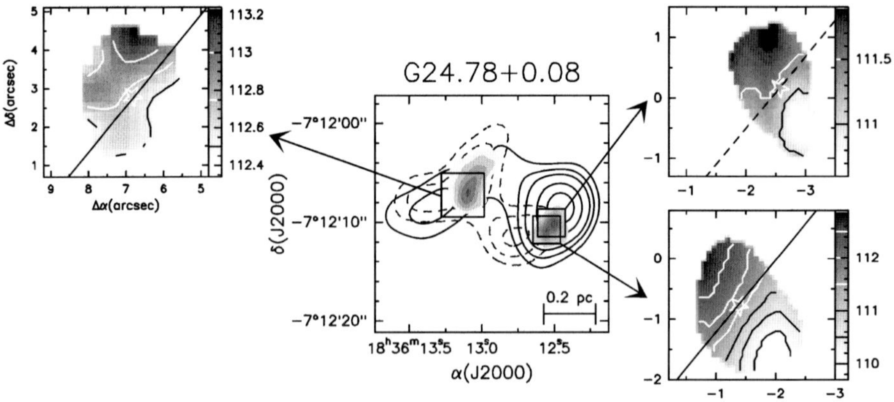

Figure 2. Toroid/outflow systems in the high-mass star forming region G24.78+0.08 (Bektrán et al., 2004). The straight lines represent the outflow axes. The stars mark the positions of the millimeter continuum peaks. Left: map of the CS(3–2) line velocity observed by Cesaroni et al. (2003) towards G24 C. The grey scale is expressed in km s^{-1}. Middle: comparison between the bipolar outflows observed by Furuya et al. (2002), the CH_3CN(12–11) line emission map towards G24 A1 and G24 A2, and the CS(3–2) line map (Furuya et al., 2002) towards G24 C. Right: map of the CH_3CN(12–11) line velocity towards G24 A1 and G24 A2.

A general conclusion that may be drawn from all these findings is that, while in B (proto)stars stable (possibly Keplerian) disks are seen, in more luminous regions only very massive, unstable toroids are found, very likely rotating about a cluster of O–B (proto)stars. We shall refer to these as "circumcluster" toroids. It must be noted that such toroids although unstable, cannot be short-lived as they have been detected in several sources: this implies that they must be continuously supplied with fresh material accreting from the surrounding molecular envelope.

The question we want to address in the following is why stable, Keplerian disks are not seen in association with O-type (proto)stars.

5. The Lifetime of Disks and Massive Stars

When studying the formation disks in early-type stars, one must take into account three competing processes: accretion onto the disk from the surrounding molecular clump; accretion from the disk onto the star or stellar cluster; ionization of the disk by the newly formed star(s). While the last two processes determine a mass loss from the disk, the first feeds the disk with new material. A detailed treatment of this problem goes obviously beyond the purpose of the present review. Here we just want to give an idea of the time needed to disperse a disk under a few arbitrary (but hopefully reasonable) assumptions. Let us suppose that the mass accretion rate onto the star, $\dot{M}_*$, is proportional to that onto the disk, $\dot{M}_{\rm acc}$, which in turn is taken proportional to the mass outflow rate, $\dot{M}_{\rm out}$. We also indicate with $\dot{M}_{\rm ion}$ the mass loss rate due to ionisation from the star(s). It is convenient to obtain an

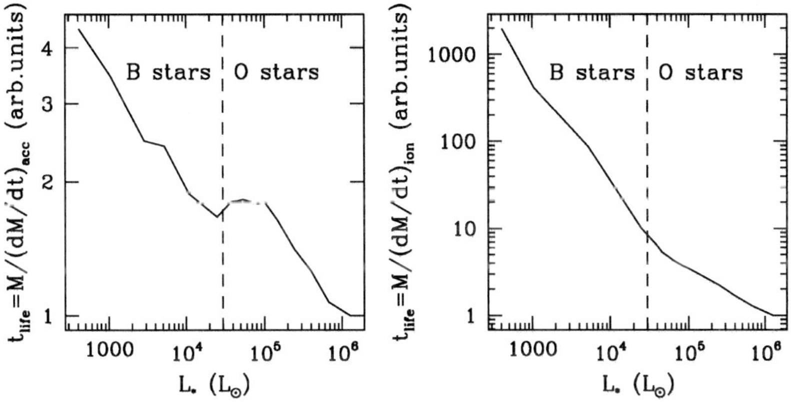

Figure 3. Dispersal timescale of disk due to accretion onto the (proto)star(s) (left panel) and ionisation by the central ZAMS star(s) (right panel).

estimate of the disk lifetime in the limits of $\dot{M}_* \gg \dot{M}_{\rm ion}$ (more suitable for the protostellar phase, where the Lyman continuum flux of the star is very small) and $\dot{M}_* \ll \dot{M}_{\rm ion}$ (typical of ZAMS early-type stars, which copiously emit UV photons). In the former case the disk lifetime is given by $t_{\rm life} = M/(\dot{M}_* - \dot{M}_{\rm acc}) \propto M_*/\dot{M}_{\rm out}$, where we have assumed that the mass of the disk is proportional to that of the (proto)star. When the star reaches the ZAMS, instead, the lifetime may be expressed as $t_{\rm life} = M/\dot{M}_{\rm ion} \propto M_*/\dot{M}_{\rm ion}$.

These expressions may be evaluated using the equation of Behrend and Maeder (2001) relating $\dot{M}_{\rm out}$ to the luminosity, and Eq. (7.10) of Hollenbach et al. (1994) for a photoevaporating disk. The results are shown in Figure 3. Note that arbitrary units are used for $t_{\rm life}$, as the proportionality constants in the relations assumed above are unknown. In both cases the lifetime is a steeply decreasing function of the luminosity: this may explain why only very massive toroids can exist in association with luminous sources, whereas in B (proto)stars much lighter disks are seen.

6. Conclusions

Observations of high-mass star forming regions have revealed the existence of powerful bipolar outflows tightly associated with rotating circumstellar, Keplerian disks or circumcluster, massive toroids. The evidence collected so far lends strong support the hypothesis expressed by various authors (e.g. Garay and Lizano, 1999) that also high-mass stars may form through non-spherical accretion as much as their low-mass counterparts, although with larger mass accretion rates.

Acknowledgement

It is a pleasure to thank Daniele Galli for critically reading the manuscript.

References

Behrend, R. and Maeder, A.: 2001, *A&A* **373**, 190.
Beltrán, M.T., Cesaroni, R., Neri, R., Codella, C., Furuya, R.S., Testi, L. and Olmi, L.: 2004, *ApJ* **601**, L187.
Bernard, J.P., Dobashi, K. and Momose, M.: 1999, *A&A* **350**, 197.
Beuther, H., Schilke, P., Sridharan, T.K., Menten, K.M., Walmsley, C.M. and Wyrowski, F.: 2002a, *A&A* **383**, 892.
Beuther, H., Schilke, P., Gueth, F., McCaughrean, M., Andersen, M., Sridharan, T.K. and Menten, K.M.: 2002b, *A&A* **387**, 931.
Beuther, H., Schilke, P. and Stanke, T.: 2003, *A&A* **408**, 601.
Bonnell, I.A. and Bate, M.R.: 2002, *MNRAS* **336**, 659.
Cesaroni, R., Olmi, L., Walmsley, C.M., Churchwell, E. and Hofner, P.: 1994, *A&A* **435**, L137.
Cesaroni, R., Felli, M., Testi, L., Walmsley, C.M. and Olmi, L.: 1997, *A&A* **325**, 725.
Cesaroni, R., Hofner, P., Walmsley, C.M. and Churchwell, E.: 1998, *A&A* **331**, 709.
Cesaroni, R., Felli, M., Jenness, T., Neri, R., Olmi, L., Robberto, M., Testi, L. and Walmsley, C.M.: 1999, *A&A* **345**, 949.
Cesaroni, R., Codella, C., Furuya, R.S. and Testi, L.: 2003, *A&A* **401**, 227.
Churchwell, E.: 2000, in: M. Livio (ed.), *Unsolved Problems in Stellar Evolution, Space Telescope Science Institute Symposium Series* **12**, 41.
Cohen, R.J., Brebner, G.C., Gasiprong, N., Hutawarakorn, B., Graham, M. and Meaburn, J.: in press, in *Proceedings of the IAU Symposium on Star Formation at High Angular Resolution, ASP Conference Series* **221**.
De Buizer, J.M.: 2003, *MNRAS* **341**, 277.
Fiebig, D., Duschl, W.J., Menten, K.M. and Tscharnuter, W.M.: 1996, *A&A* **310**, 199.
Fontani, F., Cesaroni, R., Caselli, P. and Olmi, L.: 2002, *A&A* **389**, 603.
Furuya, R.S., Cesaroni, R., Codella, C., Testi, L., Bachiller, R. and Tafalla, M.: 2002, *A&A* **390**, L1.
Garay, G. and Lizano, S.: 1999, *PASP* **111**, 1049.
Gibb, A.G., Hoare, M.G., Mundy, L.G. and Wyrowski, F.: in press, in *Proceedings of the IAU Symposium on Star Formation at High Angular Resolution, ASP Conference Series* **221**.
Greenhill, L.J., Reid, M.J., Chandler, C.J., Diamond, P.J. and Elitzur, M.: 2004, in *Proceedings of the IAU Symposium on Star Formation at High Angular Resolution, ASP Conference Series* **221**, 155.
Guilloteau, S., Dutrey, A. and Simon, M.: 1999, *A&A* **348**, 570.
Hoare, M.G., Glindemann, A. and Richichi, A.: 1996, in: H.U. Kaüfl and R. Siebenmorgen (eds.), *ESO Proceedings on the Role of Dust in the Formation of Stars*, 35.
Hofner, P., Peterson, S. and Cesaroni, R.: 1999a, *ApJ* **514**, 899.
Hofner, P., Cesaroni, R., Rodríguez, L.F. and Martí, J.: 1999b, *A&A* **345**, L43.
Hollenbach, D., Johnstone, D., Lizano, S. and Shu, F. : 1994, *ApJ* **428**, 654.
Keto, E.R., Ho, P.T.P. and Reid, M.J.: 1987, *ApJ* **323**, 117.
Keto, E.R., Ho, P.T.P. and Haschick, A.D.: 1988, *ApJ* **324**, 920.
Keto, E.R.: 2002, *ApJ* **568**, 754.
Minier, V., Booth, R.S. and Conway, J.E.: 1998, *A&A* **336**, L5.
Minier, V., Booth, R.S. and Conway, J.E.: 2000, *A&A* **362**, 1093.
Moscadelli, L., Cesaroni, R. and Rioja, M.J.: 2000, *A&A* **360**, 663.
Norris, R.P., Byleveld, S.E., Diamond, P.J., Ellingsen, S.P., Ferris, R.H., Gough, R.G., Kesteven, M.J., McCulloch, P.M., Phillips, C.J., Reynolds, J.E., Tzioumis, A.K., Takahashi, Y., Troup, E.R. and Wellington, K.J.: 1998, *ApJ* **508**, 275.
Olmi, L., Cesaroni, R., Hofner, P., Kurtz, S., Churchwell, E. and Walmsley, C.M.: 2003, *A&A* **407**, 225.
Osorio, M., Lizano, S. and D'Alessio, P.: 1999, *ApJ* **525**, 808.

Palla, F. and Stahler, S.W.: 1993, *ApJ* **418**, 414.
Phillips, C.J., Norris, R.P., Ellingsen, S.P. and McCulloch, P.M.: 1998, *MNRAS* **300**, 1131.
Sandell, G., Wright, M. and Forster, J.R.: 2003, *ApJ* **590**, L45.
Shepherd, D.S. and Churchwell, E.: 1996, *ApJ* **457**, 267.
Shepherd, D.S. and Kurtz, S.: 1999, *ApJ* **523**, 690.
Shepherd, D.S., Yu, K.C., Bally, J. and Testi, L.: 2000, *ApJ* **535**, 833.
Shepherd, D.S., Claussen, M.J. and Kurtz, S.: 1999, *ApJ* **523**, 690.
Shu, F.H., Adams, F.C. and Lizano, S.: 1987, *ARA&A* **25**, 23.
Simon, M., Dutrey, A. and Guilloteau, S.: 2000, *ApJ* **545**, 1034.
Tan, J.C. and McKee, C.F.: 2002, in *Proceedings of the ASP Conference on Hot Star Workshop III: The Earliest Stages of Massive Star Birth* **267**, 267.
Tofani, G., Felli, M., Taylor, G.B. and Hunter, T.R.: 1995, *A&AS* **112**, 299.
Torrelles, J.M., Gómez, J.F., Rodríguez, L.F., Curiel, S., Ho, P.T.P. and Garay, G.: 1996, *ApJ* **457**, L107.
Torrelles, J.M., Gomez, J.F., Rodríguez, L.F., Ho, P.T.P., Curiel, S. and Vazquez, R.: 1997, *ApJ* **598**, L115.
Torrelles, J.M., Patel, N.A., Anglada, G., Gómez, J.F., Ho, P.T.P., Lara, L., Alberdi, A., Cantó, J., Curiel, S., Garay, G. and Rodríguez, L.F.: 2003, *ApJ* **598**, L115.
Yorke, H.W.: in press, in *Proceeding of the IAU Symposium on Star Formation at High Angular Resolution, ASP Conference Series* **221**.
Yorke, H.W. and Sonnhalter, C.: 2002, *ApJ* **569**, 846.
Zhang, Q., Ho, P.T.P. and Ohashi, N.: 1998a, *ApJ* **494**, 636.
Zhang, Q., Hunter, T.R. and Sridharan, T.K.: 1998b, *ApJ* **505**, L151.
Zhang, Q., Hunter, T.R., Brand, J., Sridharan, T.K., Molinari, S., Kramer, M.A. and Cesaroni, R.: 2001, *ApJ* **552**, L167.
Zhang, Q., Hunter, T.R., Sridharan, T.K. and Ho, P.T.P.: 2002, *ApJ* **566**, 982.

MERLIN AND PUSCHINO OBSERVATIONS OF H_2O MASERS IN OUTER GALACTIC SFR S128N

A.M.S. RICHARDS[1], R.J. COHEN[1], M. CROCKER[1], E.E. LEKHT[2], E. MENDOZA[2] and V.A. SAMODUROV[3]

[1]*JBO, University of Manchester, Macclesfield, UK;*
E-mail: amsr@jb.man.ac.uk; also AVO, www.euro-vo.org.
[2]*INAOE, 72840 Tonantzintla, Puebla, México*
[3]*Astro Space Centre, LPI, Profsoyuznaya 84/32, Moscow, Russia*

(Received 16 April 2004; accepted 15 June 2004)

Abstract. We have imaged H_2O maser emission from the star-forming region S128 at milli-arcsec resolution using MERLIN, to complement 20 years of monitoring data from the Puschino radio telescope. The drift velocities of the masers and the velocity and location of a new maser region add depth to the model of two colliding CO clouds triggering collapse. Some H_2O masers appear to originate directly from this shock front. The brightest maser appears typical of a YSO jet and remains unsaturated close to peak intensity. The distribution of maser clumps has a fractal dimension $\sim$0.4; combined with analysis of drift velocity variations this suggests that the masers trace the dissipation of supersonic turbulence. The spatial distribution of velocities shows that this is in parts more structured than the Kolmogorov cascade.

Keywords: ISM: evolution, ISM: jets and outflows, masers, stars: pre-main-sequence, shock waves

1. Star Formation in S128

S128 (G97.53 + 3.19) was identified as an H II region by Sharpless (1959). It is $\sim$9 kpc from the Sun, 13–16 kpc from the Galactic Centre, and >500 pc above the Galactic plane. The main nebula is 2.5 pc in diameter, with a much smaller Ultra Compact H II region (UCHII) 2 pc to the north, S128N (Ho et al., 1981). Haschick and Ho (1985) found two H_2O maser regions (A and B in Figure 1 *Left*) bracketing the northern UCHII and suggested that star formation is occurring at the interface of two colliding CO clouds. Star formation is unusual at such a remote location with relatively low metallicity and a low molecular gas density. This is discussed further by Bohigas and Tapia (2003), who recently observed S128 using optical and IR imaging and spectroscopy. They measured the properties of 54 sources, about half of which are cluster members and show strong reddening. They infer star formation ages of $>3 \times 10^5$ and $\sim 10^6$ yr for S128N and S128, respectively, and resolve four IR nebulosities in the region in addition to two detected by *MSX*.

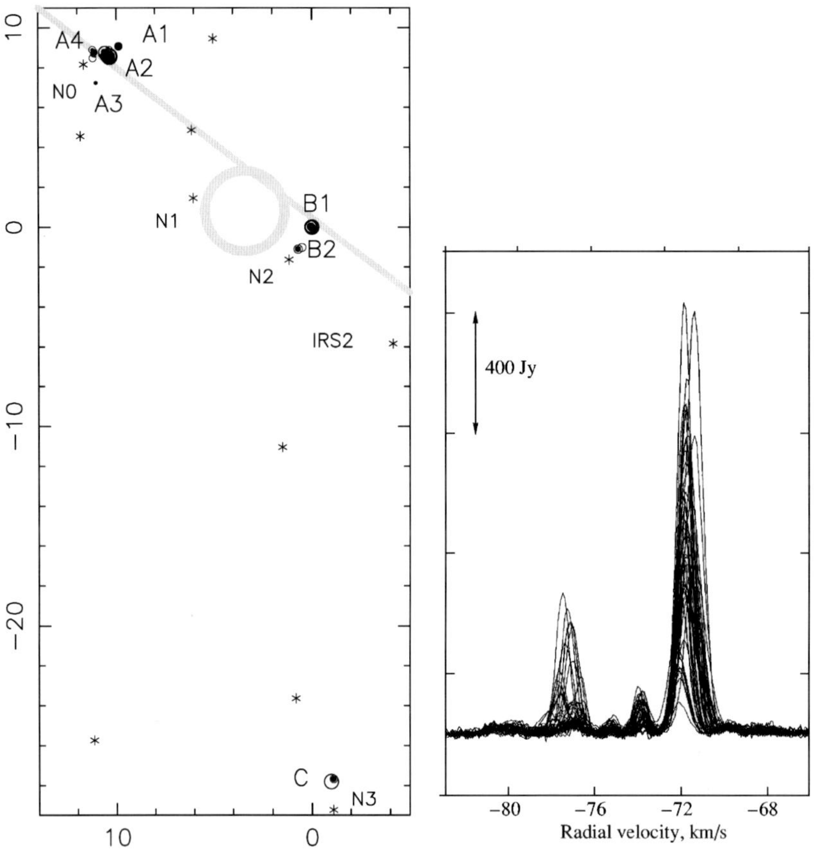

Figure 1. Left: The dark labelled circles show the positions of maser features observed in 1998 and 1999 using filled and hollow circles, respectively. The stars are positions of sources detected by Bohigas and Tapia (2003), with the IR nebulosities labelled in smaller type. The large grey circle is the approximate position of the S128N UCHII source (Ho et al., 1981) and the line is the approximate position of the CO cloud collision front (Haschick and Ho, 1985). Positions are in arcsec offset from the brightest maser. *Right*: The 22-GHz spectra from 1985 to 2001 observed towards S128 using the Puschino radio telescope.

2. Observations of 22-GHz H_2O Maser Emission

S128N has been monitored at 22 GHz since 1982 by the Puschino radio telescope (Lekht et al., 2002). Figure 1 *Right* shows some typical spectra, with an average of 3–4 peaks. We observed S128 at 22 GHz with MERLIN on January 27, 1998 and January 21, 1999, obtaining maps of maser emission with 0.1 km s^{-1} channel separation and $3\sigma_{\rm rms} \geq 40$ mJy beam^{-1} sensitivity. The absolute position of the 1998 observations was obtained by phase-referencing, giving an accuracy of (50, 65) mas in the position of the brightest maser, in region B, at $21^{\rm h}32^{\rm m}11^{\rm s}.2124\ +55°53'44''.140$ (J2000) (see Figure 1 *Left*). The 1999 observations were of longer duration under better weather conditions and give the best relative accuracy. The MERLIN beam-size is ~10 milli-arcsecond (mas) at 22 GHz, depending on data weighting. We fitted Gaussian components to each patch of maser emission, achieving mas accuracy for bright masers within ~20 arcsec of the brightest maser.

3. Maser Appearance and Origin

Figure 1 *Left* shows that the MERLIN observations detect the two regions labelled A and B by Haschick and Ho (1985) and a third region C (see Section 3.1) 30 arcsec to the south. The components occur in groups, each corresponding to the line profile of a discrete masing clump, 3–250 AU in size. About half the maser clumps occur at similar velocities and locations at both epochs and in some cases the detailed morphology is similar. Although MERLIN detected weak emission from region A across almost the whole velocity range, spatial blending contributes $<1\%$ to the Pushcino spectra peak intensities for $V_{LSR} < -70$ km s^{-1}. At the most blue-shifted end of the spectrum there is emission of comparable intensity from all three regions. The brightest spectral peaks identified with regions A and B correspond to masers separated by a light travel time ~2 yr. The MERLIN observations show that the majority of maser features show a consistent velocity drift towards the observer, of mean $\Delta V_{LSR} \sim 0.5$ km s^{-1} between 1998 and 1999. The Puschino data show a blue-shifted drift for the peak which is usually brightest, from −70.8 to −72.1 km s^{-1} between 1987 and 2001.

3.1. Newly discovered maser emission site

We find new H_2O maser emission around 68 km s^{-1} in a third region C (Figure 1 *Left*) which is >10 arcsec from the apparent CO collision front. The relatively blue-shifted maser velocities suggest that they emanate from an interaction region where the CO clouds overlap along the line of sight. This implies that the more remote CO cloud must be moving towards us more rapidly than the foreground cloud. The masers in region C peak at 7 Jy/channel in 1998 and 10 Jy/channel in 1999, which would have been detected by Haschick and Ho (1985), but the spectra of Lekht et al. (2002) show that emission at these velocities is intermittent. Region C is approximately mid-way between the two H II regions and close to IR nebulosity N3 (Bohigas and Tapia, 2003), just as regions A and B are within ~1 arcsec of the IR nebulosities labelled N0 and N2.

3.2. Maser emission from extended shock fronts

A number of maser clumps in regions A and B are elongated suggesting that they emanate from a shock front approximately perpendicular to the plane of the

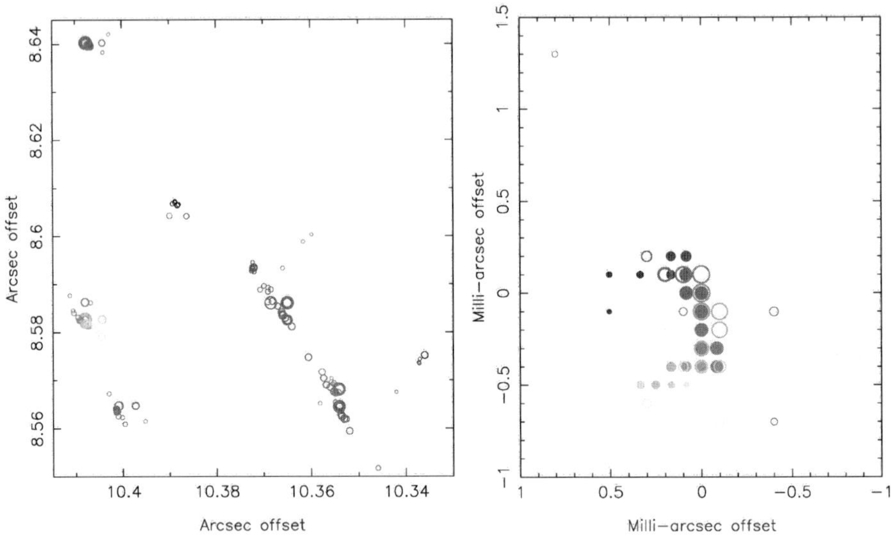

Figure 2. The filled and hollow symbols show the positions of maser components observed in 1998 and 1998, respectively. The diameter is proportional to log flux density and the grey scale shows the relative V_{LSR}, darker implying more red-shifted. The two plots are on different scales; all positions are with respect to the brightest maser. *Left*: Masers in features observed at both epochs in region A2. The velocity range is from −82.5 to −72.5 km s^{-1}. *Right*: The brightest maser feature at both epochs, in region B1. The velocity range is from −73.5 to −69.0 km s^{-1}.

sky. The largest such region, A2 (Figure 2 *Left*), covered 15 km s^{-1} in V_{LSR} in 1998 and almost 1000 AU in length, with a smaller parallel region 350 AU away. There is no systematic large-scale velocity gradient although individual maser clumps have local gradients. The direction of maser elongation is mostly approximately NE-SW, similar to the CO cloud collision front. We suggest that at least some of these H_2O masers are excited by the cloud collision, not by any specific YSO.

At the distance of S128, the minimum detectable proper motion during the 1 year separation of the MERLIN epochs would be 2 mas (∼100 km s^{-1}). Apparent motions of this magnitude are seen in a few clumps, which can be unambiguously matched at both epochs. The change in position is parallel to the direction of elongation of the maser clumps. This speed is feasible for YSO jets and in W75(N) masers are seen with similar morphology associated with a radio-continuum jet (Torrelles et al. (2003) and these proceedings). The maximum observed V_{LSR} range is <20 km s^{-1}. It is possible that jets propagating in the plane of the sky produce favourable maser beaming. The consistent orientation parallel to the CO cloud collision front, however, suggests that the masers may be associated with disturbances travelling along the shock front with a pattern speed of ∼100 km s^{-1}, not bulk motion of the masing gas.

3.3. Variability of a Maser Associated with a Jet

Lekht et al. (2002) found that over an 18-month period the brightest maser feature, around -71.8 km s^{-1}, showed a correlation between variations in line width ΔV and peak flux density F_0 approximated by $\Delta V^{-2} = C \ln F_0$. C describes a curve which is steeper during brightening, when the line narrows more rapidly than the rebroadenning rate when dimming. This suggests that the maser remains unsaturated as it brightens and that the effective gain length falls off more rapidly than the pump rate as it dims. The MERLIN images (taken at the start of the rise in intensity and during the decline) show that this feature (in B1, Figure 2 *Right*) is a very compact arc of radius 3 AU, typical of masers excited at the working surface of a YSO jet (Torrelles, these proceedings; Brand et al., 2003). The long variability timescale relative to the small maser clump size suggests that mechanical rather than radiative processes are directly responsible. Maser emission is excited close to the shock front, but disrupted by turbulence.

4. Maser Distribution and Turbulence

Lekht et al. (1999) present a method for the analysis of variations in the drift velocities of spectral maser features with time. Applied to S128, they obtain Strouhal numbers $\ll$1, indicating either incompressible turbulence or chaotically propagating shock waves.

We use the method of Strelnitski et al. (2002) to look for indications of fractal behaviour in the spatial clustering of masers. The maser features are mostly well separated on the sky so that their volume density is directly related to their surface density σ. For a pair separation r, $\sigma \propto r^{d-d_0}$ where d_0 is the dimension of the supporting space, in this case 2, and d is the fractal dimension. Differentiation gives

$$d = \frac{d(\log \sigma)}{d(\log r)} + d_0 \qquad (1)$$

We calculated σ as a function of logarithmically spaced increments in r for all feature clumps at both epochs, and for the distribution of components within features in regions A and B in 1999 where the best position accuracy was obtained. The results, plotted in Figure 3, show that within features $d(\log \sigma)/d(\log r) \sim -2.2 \pm 0.2$ so $d \sim 0$; the components are uniformly distributed within the limits of our accuracy. The distribution of features, however, has a mean $d(\log \sigma)/d(\log r) = -1.62 \pm 0.06$ giving $d = 0.38 \pm 0.06$. Incompressible turbulence has a fractal dimension of $\sim$2.6 (Mandelbrot, 1982) so the low value of d for S128 show that the H_2O maser features mostly trace the chaotic dissipation of supersonic turbulence.

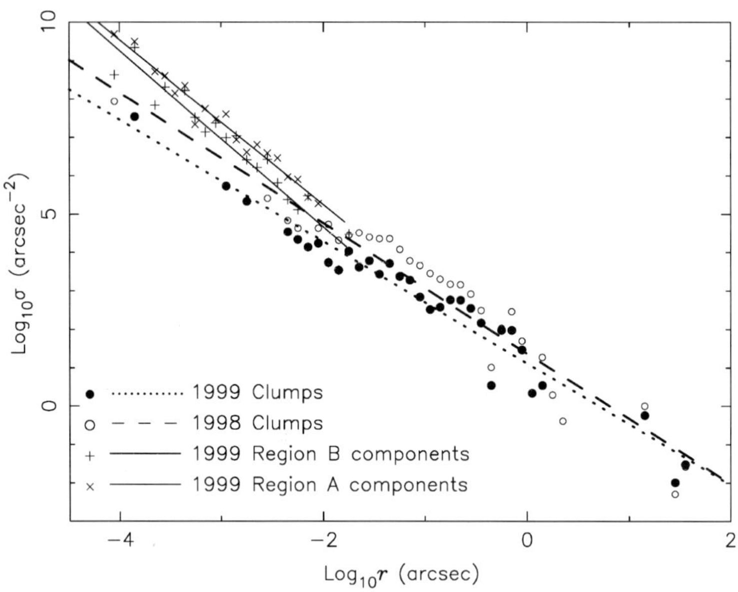

Figure 3. The surface density σ of pairs of masers with separation in the range r.

Strelnitski et al. (2002) also use the mean velocity separation as function of pair separation, $D_\alpha(\delta \log r) = (\sum |\delta V_{\rm LSR}|^\alpha)/N$, where in each logarithmically spaced interval of angular separation $\delta \log r$, there are N pairs, each with velocity separation $\delta V_{\rm LSR}$. In the case of Kolmogorov turbulence, for $\alpha = 1, 2, 3$, $d \log D_\alpha / d \log r = q = 1/3, 2/3, 1$, respectively. We found that the relevant quantities for the H_2O maser features in S128 were not tightly correlated and obtained $q = 0.10 \pm 0.03$, 0.54 ± 0.08, 0.96 ± 0.16. The large scatter is consistent with the non-random distribution of some of the features, especially in region A where they may lie along the CO cloud collision interface, see Section 3.2.

5. Conclusions

The position and velocity of the newly detected maser region C suggests that the CO cloud collision is occurring at an oblique angle to the line of site. Combined with evidence from maser drift velocities, this suggests that the more remote CO cloud is pushing the collision region towards us. The star-formation ages deduced by Bohigas and Tapia (2003) suggest the disturbance propagated from south to north. The presence of masers at C suggests that the more recent star formation around S128N extends at least halfway towards the older southern region. It has previously been suggested that region A and B masers may be associated with a shock front around the UCHII region. Some of these clumps are elongated over hundreds of AU and, along with region C, could be associated directly with the CO cloud collision rather than with any specific YSO.

Acknowledgements

MERLIN is the Multi Element Radio Linked Interferometer Network, operated by the University of Manchester on behalf of PPARC. AMSR gratefully acknowledges financial support from the UK Royal Society, the hospitality of INOAEP. We thank the organisers of Molecules 2004 for a very interesting conference.

References

Bohigas, J. and Tapia, M.: 2003, *AJ* **126**, 1861.
Brand, J., Cesaroni, R., Comoretto, G., Felli, M., Palagi, F., Palla, F. and Valdettaro, R.: 2003, *A&A* **407**, 573.
Haschick, A.D. and Ho, P.T.P.: 1985, *ApJ* **292**, 200.
Ho, P.T.P., Haschick, A.D. and Israel, F.P.: 1981, *ApJ* **243**, 526.
Mandelbrot, B.B.: 1982, *The Fractal Geometry of Nature*, Freeman, New York.
Lekht, E.E., Mendoza-Torres, J.E. and Berulis, I.I.: 2002, *ARep.* **46**, 57.
Lekht, E.E., Mendoza-Torres, J.E. and Silantév, N.A.: 1999, *ARep.* **43**, 209.
Sharpless, S.: 1959, *ApJS* **4**, 257.
Strelnitski, V., Alexander, J., Gezari, S., Holder, B.P., Moran, J.M. and Reid, M.J.: 2002, *ApJ* **581**, 1180.
Torrelles, J.M., Patel, N.A., Anglada, G., Gómez, J.F., Ho, P.T.P., Lara, L., Alberdi, A., Cantó, J., Curiel, S., Garay, G. and Rodríguez, L.F.: 2003, *ApJ* **598**, L115.

OBSERVING MAGNETIC FIELDS IN STAR-FORMING REGIONS

JIM COHEN

The University of Manchester, Jodrell Bank Observatory, Macclesfield, Cheshire, UK;
E-mail: rjc@jb.man.ac.uk

(Received 16 April 2004; accepted 15 June 2004)

Abstract. Magnetic field strengths and directions can be estimated using many different observational techniques that span a wide range of wavelengths. Each observational method favours different regimes of scale size, density and other physical conditions. The available techniques and their ranges of applicability are briefly described and the current status of observations is reviewed, with particular emphasis on high-resolution observations of star-forming regions.

Keywords: magnetic fields, star-forming regions, Zeeman effect, masers, polarization

1. Introduction

In 1969 I was a vacation student at Mount Stromlo and Siding Springs Observatories, helping Don Mathewson to measure the polarization of starlight. The fact that there was an interstellar magnetic field intrigued me, and so did the fact that starlight carried its imprint. My next encounter with cosmic magnetic fields came in 1983 when I began to study masers and bipolar molecular outflows from young stars. I learned that the prototype outflow, from the source L1551, has its CO emission lobes parallel to the interstellar polarization vectors in that part of the sky (Snell et al., 1980). What a connection! Checking the literature I found that other outflows were aligned with the local interstellar magnetic field. Furthermore the magnetic field measured using hydroxyl (OH) masers was strong enough to affect the dynamics of the outflow (Cohen et al., 1984).

Magnetic fields are also important in early stages of star-formation. The balance between magnetic, gravitational and turbulent energy is critical for the stability of molecular clouds (e.g. Zweibel and McKee, 1995). Gravitational collapse of molecular clouds proceeds preferentially along the magnetic field lines, giving rise to large rotating disc or torus structures orthogonal to the magnetic field. The magnetic field can play a key role in halting the collapse and in transferring angular momentum. It is natural to expect that there may be polarization signatures for the different evolutionary stages of a young stellar object (Vallée and Bastien, 2000). Vallée and Bastien present a sequence of possible magnetic field configurations associated with different evolutionary classes of object (Classes I–V). The basic idea is that the magnetic field starts uniform and is progressively distorted by gravitational

Astrophysics and Space Science **295:** 27–36, 2004.

Chuss et al. (2003) measured the 350-μm continuum polarization of the galactic centre using the Hertz polarimeter on the Caltech Submillimeter Observatory (CSO). They found a two-phase structure to B. Where the gas density is high B runs parallel to the galactic plane (toroidally about the centre). Where the gas density is lower B runs perpendicular to the plane, as in the radio filament mapped by Lang et al. (1999). Chuss et al. (2003) argue that cloud collapse along field lines can produce a toroidal B field inside the dense molecular gas whilst leaving B poloidal outside the molecular clouds. Their model resembles the Uchida and Shibata (1985) model for a collapsing molecular cloud. Chuss et al. argue that magnetic reconnection provides the energy to power the non-thermal radio filaments, although other interpretations cannot be excluded.

In all these cases the polarization gives information only about the direction of the magnetic field. If we want to estimate the magnitude of B then we need to make assumptions about the dust.

Crutcher et al. (2004), Vallée et al. (2003) and others have used the randomness in the polarization vectors to estimate B, based on an argument due to Chandrasekhar and Fermi (1953) concerning magneto-hydrodynamic waves. According to these authors we should be able to estimate the component of B in the plane of the sky, $B \sin\theta$, from the rms deviation of polarization angles from the mean, assuming that MHD waves are responsible for the deviations: $B \sin\theta \propto n^{1/2} \delta V \delta\phi^{-1}$, where n is the gas density, δV is the turbulent velocity and $\delta\phi$ is the rms deviation in the polarization angle ϕ. The technique has previously been applied to far infrared photometry by Chrysostomou et al. (1994) and Gonatas et al. (1994). Using SCUBA measurements at 850 μm Crutcher et al. estimate magnetic fields of 80 μG in L183 and 140 μG in L1544, while Vallée et al. estimate a field of 150 μG in the Bok Globule CB 068.

3. Zeeman Measurements

The most direct measurements of the magnetic field come from the Zeeman effect. An electron in a magnetic field B processes at the Larmor frequency $\omega_L = \frac{eB}{2m_e}$ (in SI units), where e is the electronic charge and m_e is the mass of the electron. In the classical Zeeman effect shown in Figure 1 a spectral line is split into three polarized components at frequencies ω_0, ω_0+ω_L and ω_0-ω_L. Looking along the field lines we see a pair of oppositely circularly polarized σ-components; looking perpendicular to B we see three linearly polarized components, the central π-component parallel to B and the two outer σ-components perpendicular to B. In general at an angle θ we see a linearly polarized π-component and two elliptically polarized σ-components.

If the Zeeman splitting is strong enough to separate the three line components then we get a direct measurement of B. If however the Zeeman components are blended then we measure $B \cos\theta$.

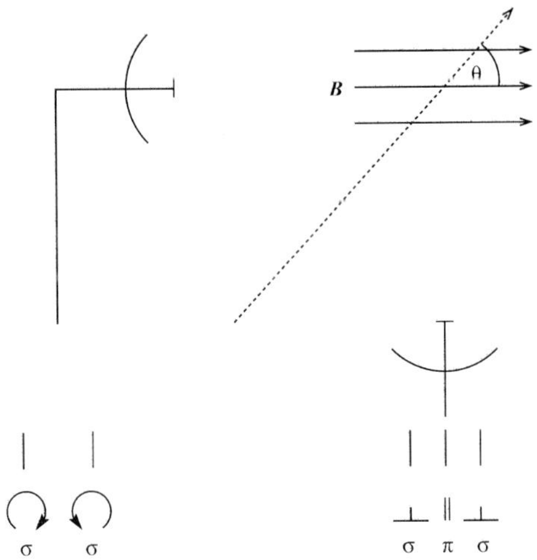

Figure 1. The classical Zeeman effect. If the Zeeman components are blended then observations at an angle θ to the magnetic field give $B\cos\theta$, whereas if the components are separated then observations yield B directly.

3.1. Thermal Lines

The classical Zeeman effect is a sufficient basis to understand Zeeman splitting of the 1420-MHz line of neutral atomic hydrogen. The splitting is weak, so σ-components dominate and we measure only weak circular polarization, with the Stokes parameter $V = zB\cos\theta\frac{dI}{d\nu}$, where z is the splitting factor and $I(\nu)$ is the line profile. Notice that whatever the magnetic field, the Stokes V-profile is unchanged. Increasing B simply increases the amplitude. Instrumental issues and sensitivity limit the usefulness of 1420-MHz Zeeman studies to absorption measurements and fields exceeding $\sim$10 μG.

Absorption line studies by Brogan and Troland (2001) yielded $B\cos\theta$ values of up to $\sim$750 μG in the star-forming region M17. $B\cos\theta$ increases towards the Western edge of the source, where $B\sin\theta$ (traced by 100-μm polarization) decreases. Either the magnetic field is bending around the edge of the HII region, or the dust properties are being modified by the HII region.

Molecular line Zeeman studies require some understanding of the quantum Zeeman effect. In a magnetic field the upper and lower states of a molecular transition each split into $2F + 1$ levels, where F is the total quantum number. The allowed transitions are $\Delta m = \pm 1$ (σ-components) and $\Delta m = 0$ (π-components), where m is the magnetic quantum number. Because the upper and lower states have different splitting factors in general, the complete Zeeman can be complex.

For the mainlines of the OH ground state, at 1665 and 1667 MHz, the upper and lower states split by equal amounts for each line, so the Zeeman patterns are simple and straightforward to interpret. Sarma et al. (2000) studied OH mainline absorption against continuum sources in NGC6334 and measured fields of $\sim$200 μG. Virial estimates suggest that the magnetic fields are of the right order to support the molecular clouds associated with these sources against gravitational collapse.

A remarkable achievement was the detection of the Zeeman effect in thermal emission from the OH 1665 and 1667-MHz lines (Crutcher and Troland, 2000), using Arecibo. This technique has the potential to probe the magnetic field right across entire molecular clouds, not just in those regions where there are continuum sources. OH thermal emission is known over the whole Taurus complex (Wouterloot and Habing, 1985), for example, a region tens of degrees across.

The presence of many hyperfine components usually complicates the Zeeman pattern but complication can sometimes give new possibilities. The CN 1-0 transition near 113 GHz has nine hyperfine components that are well separated in velocity. Crutcher et al. argued that these lines provide the best opportunity to measure magnetic fields in clouds with densities of 10^5–$10^6\,\mathrm{cm}^{-3}$. Because the different components couple differently to B (one even goes the other way) the effect of systematic and instrumental errors can be reduced. A clear detection of the Zeeman effect in these CN hyperfines was made using the Pico Veleta telescope (Crutcher et al., 1999). By fitting simultaneously to the seven major components Crutcher et al. found $B\cos\theta = 0.36 \pm 0.08\,\mathrm{mG}$ in OMC1 and similar values in DR21(OH) and M17.

Crutcher (1999) summarized the then available Zeeman data on thermal lines and concluded that the magnetic and kinetic energies are approximately equal, with the magnetic field scaling with gas density as $B \propto \rho^{0.47}$. This result is consistent with ambipolar diffusion, but it can also be interpreted as saying that $v_{\mathrm{Alfvén}} = 0.7\,\mathrm{km\,s}^{-1}$. The maser measurements described in the next section extend the range of validity of this relationship by four orders of magnitude.

Linear polarization of molecular spectra can also arise through a quantum mechanical effect (Goldreich and Kylafis, 1981). The linear polarization is either parallel or perpendicular to the magnetic field direction. The effect was first detected in CO emission from the galactic centre (Greaves et al., 1999) and has since been applied to star-forming regions by Lai et al. (2003) and others. Although the polarization is weak, this technique has the advantage of being potentially applicable to any molecular line in any region of a molecular cloud.

3.2. MASER LINES

Masers provide the only direct measurement of B at subarcsecond resolution, probing densities from $\sim 10^5$–10^8 cm^{-3} for OH, to 10^{10} cm^{-3} for H_2O. The Zeeman effect gives only part of the polarization story however, since masers are nonlinear. There are other effects to consider, including magnetic beaming, saturation and blending of maser lines.

As in the case of thermal lines, the interpretation is much cleaner if the magnetic field splits the Zeeman components by more than their thermal linewidths. This is the case for the most commonly observed OH masers at 1665 and 1667 MHz. The maser amplification often introduces an asymmetry into the amplitudes of the Zeeman features, since all molecules contributing to a particular maser beam must have the same resonant frequency. The magnetic shift of frequency is equivalent to a velocity shift. Therefore along any given line-of-sight different molecules contribute to beams of right-hand circular (RHC) and left-hand circular (LHC) polarization. Hence the gains of the RHC and LHC beams are different in general, and one or other Zeeman component dominates. Consequently complete Zeeman patterns are rarely seen. Furthermore, masers amplify most strongly along the magnetic field lines (Gray and Field, 1995), a phenomenon termed "magnetic beaming". As the maser saturates the stronger beam (along B) increasingly dominates, so circular polarization and σ-components dominate. Data for OH 1665-MHz masers in W75N

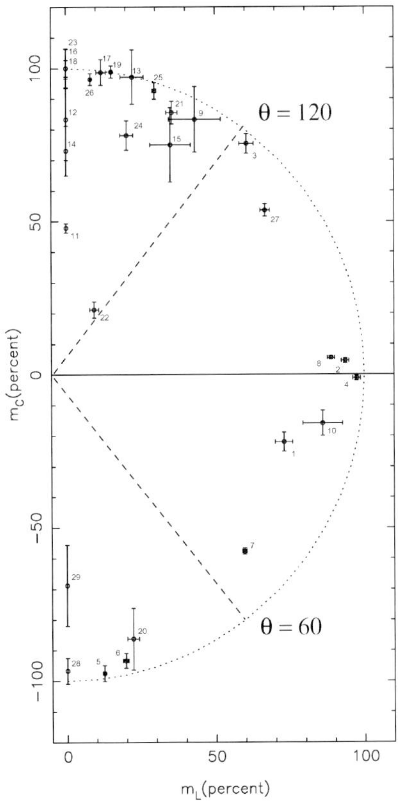

Figure 2. Circular polarization as a function of linear polarization for OH 1665-MHz masers in W75N, measured with MERLIN (Hutawarakorn et al., 2002). Because of magnetic beaming, OH masers amplify most strongly along the line-of-sight, corresponding to $\theta \sim 0$ or $\theta \sim 180$ degrees.

shown in Figure 2 illustrate the beaming effect. We see mostly σ-components with strong circular or elliptical polarization corresponding to θ close to the line-of-sight, and very few π-components corresponding to $\theta \simeq 90°$.

Because of magnetic beaming it is reasonable to assume that elliptically polarized components are σ-components. If that is the case then we can deduce θ, the angle of B to the line-of-sight, from the degree of circular polarization, since if the masers are not saturated then $V = -2I\cos\theta/(1+\cos^2\theta)$ as for thermal emission. From the linear polarization we know the projected angle of B on the plane of the sky, so we can in fact determine direction of the B vector in three dimensions (Garcia-Barreto et al., 1988). If we are lucky enough to see a Zeeman pair then splitting gives us the magnitude of B, and we can determine the three-dimensional B vector.

MERLIN observations of OH masers associated with bipolar outflows have revealed maser tori or discs at the centre of the flows, with scale ~2000 AU (Hutawarakorn and Cohen, 1999, 2003; Hutawarakorn et al., 2002). The direction of B reverses on opposite sides of the torus, suggesting that rotation has wound up the field. Detailed modelling of the OH masers in W75N (Gray et al., 2003) supports the twisted magnetic field model proposed by Uchida and Shibata (1985). The NGC7538 region contains three bipolar outflows each with OH masers (IRS1, IRS9 and IRS11) but with very different degrees of polarization. The same behaviour is seen in SCUBA data on scales 100 times larger: IRS11 has well-organized B vectors aligned with the outflow, while IRS1 has less polarization and less organized B vectors (Momose et al., 2001). It is suggested that the polarization properties are age related, with younger sources being more highly organized, having stronger polarization and greater alignment with the outflow direction.

Excited OH masers at 6.0 GHz have Zeeman splitting comparable to their linewidths, so that Zeeman pairs are common. Caswell (2003) carried out an extensive search for 6.0-GHz OH masers and found many cases of matching Zeeman pairs at 6035 and 6030 MHz, from which reliable B values can be deduced even without interferometry.

Blending of maser lines (including hyperfine components) is difficult to deal with. The greater the blending and the greater the degree of saturation, the bigger have been the disagreements over how to interpret the data. A summary of the current theoretical situation is given by Gray (2003). The strong H_2O masers at 22 GHz show weak polarization, both circular (Sarma et al., 2002) and linear (Imai et al., 2003), but the interpretation is complicated by the presence of blended hyperfine components, and will not be considered further here. See however Imai, these proceedings. Likewise polarization of SiO and CH_3OH masers will not be discussed here, despite exciting new observational results from Plambeck et al. (2003) and Gaylard and Goedhart (these proceedings).

4. Conclusions

Sensitivity limits three-dimensional magnetic field studies at present, but the potential to probe a wide range of densities up to at least $\sim 10^{10}$ cm^{-3} has been demonstrated. Major new facilities such as ALMA should bridge the gap between masers and other molecular lines, enabling us finally to see the magnetic field structure of whole molecular cloud complexes, on angular scales from degrees down to milliarcseconds.

Key questions for the future include the treatment of overlapping hyperfine components, and the issues that got me started in this field in the first place, the relation of the magnetic field in star-forming regions to the general galactic field and to the evolution of the young stars.

Acknowledgements

I thank Malcolm Gray and Busaba Kramer (née Hutawakorn) for interesting discussions.

References

Brogan, C.L. and Troland, T.H.: 2001, *ApJ* **560**, 821.
Brown, J.C., Taylor, A.R., Wielebinski, R. and Müller, P.: 2003, *ApJ* **592**, L29.
Brown, J.C. and Taylor, A.R.: 2001, *ApJ* **563**, L31.
Caswell, J.L.: 2003, *MNRAS* **341**, 551.
Chandrasekhar, S. and Fermi, E.: 1953, *ApJ* **118**, 113.
Chrysostomou, A., Hough, J.H., Burton, M.G. and Tamura, M.: 1994, *MNRAS* **268**, 325.
Chuss, D.T., Davidson, J.A., Dotson, J.L., Dowell, C.D., Hildebrand, R.H., Novak, G. and Vaillancourt, J.E.: 2003, *ApJ* **599**, 1116.
Cohen, R.J., Rowland, P.R. and Blair, M.M.: 1984, *MNRAS* **210**, 424.
Crutcher, R.M.: 1999, *ApJ* **520**, 706.
Crutcher, R.M., Nutter, D.J., Ward-Thompson, D. and Kirk, J.M.: 2004, *ApJ* **600**, 279.
Crutcher, R.M. and Troland, T.H.: 2000, *ApJ* **537**, L139.
Crutcher, R.M., Troland, T.H., Lazareff, B. and Kazès, I.: 1996, *ApJ* **456**, 217.
Crutcher, R.M., Troland, T.H., Lazareff, B., Paubert G. and Kazès, I.: 1999, *ApJ* **514**, L121.
Fischer, O., Henning, Th. and Yorke, H.W.: 1994, *A&A* **284**, 187.
Garcia-Barreto, J.A., Burke, B.F., Reid, M.J., Moran, J.M., Haschick, A.D. and Schilizzi, R.T.: 1988, *ApJ* **326**, 954.
Goldreich, P. and Kylafis, N.D.: 1981, *ApJ* **243**, L75.
Gonatas, D.P., Engargiola, G.A., Hildebrand, R.H., Platt, S.R., Wu, X.D., Davidson, J.A., Novak, G., Aitken, D.K. and Smith, C.: 1994, *ApJ* **357**, 132.
Gray, M.D.: 2003, *MNRAS* **343**, L33.
Gray, M.D. and Field, D.: 1995, *A&A* **298**, 243.
Gray, M.D., Hutawarakorn, B. and Cohen, R.J.: 2003, *MNRAS* **343**, 1067.

the results for 1665- and 1667-MHz transitions will be described in a separate paper.

The second target W75N showed a spectrum at 4765 MHz with three features within a velocity range from 9.5 to 10.6 km s^{-1} (Figure 2). The 4765-MHz components are located on the S–W edge of the H II region mapped in the radio continuum and labelled as Ba by Hunter et al. (1994) or VLA 1 by Torrelles et al. (1997) (Figure 2). Similarly to the 4765-MHz emission from Cep A they also appeared as three groups of spots but were aligned along the N–S direction (Figure 2). They also showed a velocity gradient from the north (blue-shifted) to the south (red-shifted). The position of the 4765-MHz maser structure coincides to within 60 mas with the position of the 1720-MHz Zeeman pair Z_6 reported by Hutawarakorn et al. (2002), while the central velocity of the 1720-MHz Zeeman pair is very close (0.3 km s^{-1}) to the velocity of the most blue-shifted feature of the 4765-MHz profile. The position of the 4765-MHz maser centre observed with MERLIN agrees well with that measured with an accuracy of about 1$''$ using VLA in 1984 (Palmer et al., 2004). However, their spectrum is very different from ours.

4. Discussion and Conclusions

The 4765- and 1720-MHz masers from Cep A and W75N appeared within the same velocity ranges and coincided in the sky well within the position uncertainties indicating their coexistence in the space. Especially in Cep A the result is convincing as the observations of those two transitions were carried out at two epochs only 2.5 months apart. The extrapolation for 4765-MHz flux density from Cep A using the fitting from Section 3 gives 45 mJy on May 20, 1999, the epoch of the 1720-MHz observation. This implies a ratio of $\sim$25 ($\pm$5) between the peak flux densities at 1720 and 4765 MHz. Such an intensity ratio can be explained using the model by Gray et al. (1991). They proposed low density regime: n_{H_2} of 4×10^6 cm^{-3}, n_{OH} of 2×10^2 cm^{-3} and a kinetic temperature range from 100 to 200 K, where there is no overlap between the pump line transitions for the different maser lines, assuming pumping by far infrared photons. In case of a large-scale velocity field of up to 1.7 km s^{-1} across the masing region, line overlap allows the 1720-MHz maser to survive to higher densities with n_{H_2} up to a few $\times10^7$ cm^{-3}. Their model also predicts the absence of the main line ground-state OH masers and the other two 4.7-GHz lines. This is fully consistent with our observations; no emission was detected at 1665-, 1667-, 4660- or 4750-MHz, in a projected area 2$''$ $\times$ 2$''$ centred at $-$14.2 km s^{-1} feature of the 4765-MHz emission.

The survey by Caswell (2004) showed OH 1720-MHz and CH_3OH 6.7-GHz correlation. In Cep A the 12.2-GHz and in W75N the 6.7-GHz methanol masers were mapped (Minier et al., 2000) but their positions are a few arcsec far away from 1720-/4765-MHz regions. Concluding, in both sources we found the association of

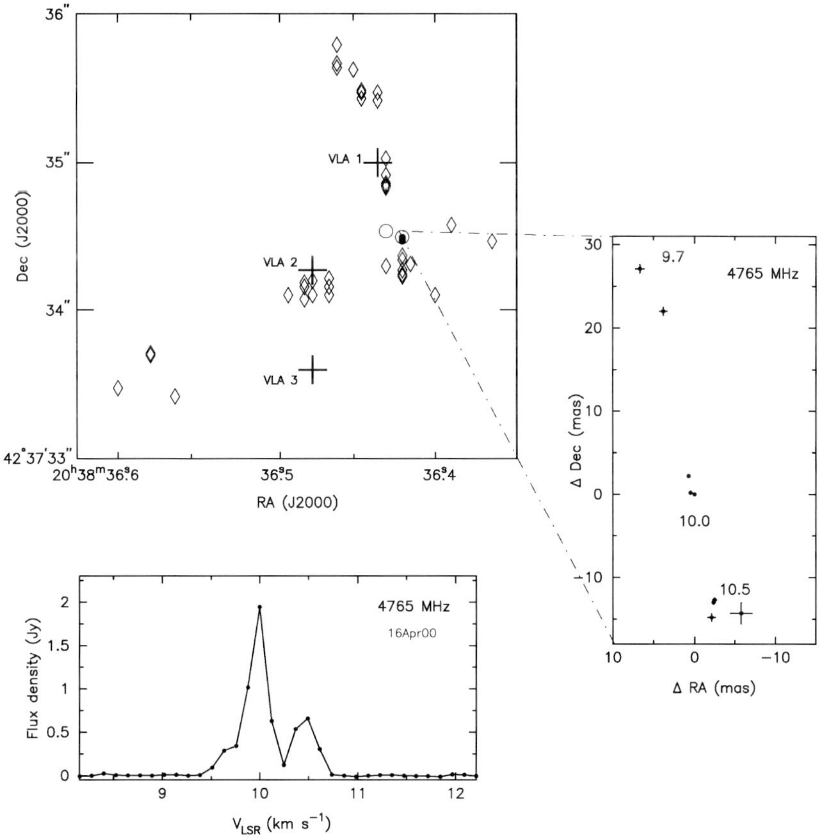

Figure 2. Top: the distribution of 1665-, 1720- (Hutawarakorn et al., 2002) and 4765-MHz OH masers (this paper) in W75N marked as diamonds, open circles and dots, respectively. The crosses and labels indicate the centres of H II regions taken from Torrelles et al. (1997). Relative positions of OH 4765-MHz masers is presented on the enlarged region. The central velocity V_{LSR}(km s^{-1}) of each group and relative position errors of components are given. *Bottom:* the total flux density spectrum of the OH 4765-MHz emission from W75N taken with MERLIN on April 16, 2000.

1720- and 4765-MHz OH masers that strongly supports the model by Gray et al. (1991).

Acknowledgements

MERLIN is a national facility operated by the University of Manchester at Jodrell Bank on behalf of PPARC. AN acknowledges a fellowship funded by the EU under the Marie Curie Training Site programme and the travel and accommodation support from the LOC of the European Workshop 2004 on Astronomical Molecules. The work was also supported by grant 2P03D01122 of the Polish State Committee for Scientific Research.

References

Argon, A.L., Reid, M.J. and Menten, K.M.: 2000, *ApJS* **129**,159.
Baart, E.E., Cohen, R.J., Davies, R.D., Norris, R.P. and Rowland, P.R.: 1986, *MNRAS* **219**, 145.
Blaauw, A., Hiltner, W.A. and Johnson, H.L.: 1959, *ApJ* **130**, 69.
Caswell, J.L.: 2004, *MNRAS* **349**, 99.
Cohen, R.J., Rowland, P.R. and Blair, M.M.: 1984, *MNRAS* **210**, 425.
Diamond, P.J., Garrington, S.T., Gunn, A.G., Leahy, J.P., McDonald, A., Muxlow, T.W.B., Richards, A.M.S. and Thomasson, P.: 2003, *MERLIN User Guide*, version 3.
Dickel, J.R., Dickel, H.R. and Wilson, W.J.: 1978, *ApJ* **223**, 840.
Hughes, V.A. and Wouterloot, J.G.A.: 1984, *ApJ* **276**, 204.
Hunter, T.R., Taylor, G.B., Felli, M. and Tofani G.: 1994, *A&A* **284**, 215.
Hutawarakorn, B., Cohen, R.J. and Brebner, G.C.: 2002, *MNRAS* **330**, 349.
Garay, G., Ramirez, S., Rodriguez, L.F., Curiel, S. and Torrelles, J.M.: 1996, *ApJ* **459**, 193.
Gray, M.D., Doel, R.C. and Field, D.: 1991, *MNRAS* **252**, 30.
Gray, M.D., Field, D. and Doel, R.C.: 1992, *A&A* **262**, 555.
Minier, V., Booth, R.S. and Conway, J.E.: 2000, *A&A* **362**, 1093.
Niezurawska, A., Szymczak, M., Cohen, R.J. and Richards, A.M.S.: 2004, *MNRAS* **350**, 1409.
Palmer, P., Goss, W.M. and Whiteoak, J.B.: 2004, *MNRAS* **347**, 1164.
Szymczak, M., Kus, A.J. and Hrynek, G.: 2000, *MNRAS* **312**, 211.
Torrelles, J.M., Gomez, J.F., Rodriguez, L.F., Ho, P.T.P., Curiel, S. and Vazquez, R.: 1997, *ApJ* **489**, 744.
Torrelles, J.M., Gomez, J.F., Garay, G., Rodriguez, L.F., Curiel, S., Cohen, R.J. and Ho, P.T.P.: 1998, *ApJ* **509**, 262.

MAGNETIC FIELDS IN STAR-FORMING REGIONS: THEORETICAL ASPECTS

DANIELE GALLI
INAF-Osservatorio Astrofisico di Arcetri, Largo E. Fermi 5, Firenze, Italy;
E-mail: galli@arcetri.astro.it

(Received 16 April 2004; accepted 15 June 2004)

Abstract. We review the basic theoretical elements leading to our current understanding of the role of magnetic fields in the process of star formation. In particular, we concentrate on: (i) the relevance of the mass-to-flux ratio for the stability of molecular clouds; (ii) the consequences of magnetic flux leakage for the evolution of cloud cores; (iii) the phase of anisotropic dynamical collapse following the formation of strongly peaked density distributions; (iv) the mechanism of magnetic braking as a possible solution to the angular momentum problem in star formation.

Keywords: magnetic fields, molecular clouds, star formation

1. Introduction

Observational evidence on the existence and the physical properties of magnetic fields in molecular clouds has been obtained mainly by two methods: (i) measurements of the Zeeman splitting of hyperfine transitions of atoms and molecules, and (ii) measurements of polarization of the thermal radiation emitted at millimetre/submillimetre wavelengths by magnetically aligned aspherical dust grains (see e.g. Cohen, 2004, these proceedings, for a detailed discussion of these and others observational techniques).

Here, it is sufficient to recall that the Zeeman effect in thermal (H, OH, and CN) and maser (OH and H_2O) lines is the most direct way of measuring the integrated line-of-sight component of the magnetic field: HI Zeeman observations sample gas densities from $n \sim 10\,\mathrm{cm}^{-3}$ to $10^2\,\mathrm{cm}^{-3}$, OH Zeeman observations cover the range from $n \sim 10^3\,\mathrm{cm}^{-3}$ to $10^4\,\mathrm{cm}^{-3}$, whereas OH and H_2O masers probe the field strength in very dense gas from $n \sim 10^6\,\mathrm{cm}^{-3}$ up to $10^{11}\,\mathrm{cm}^{-3}$. However, in the cloud cores where star formation takes place, the gas is molecular and contains little HI, whereas the abundance of the highly reactive OH radical decreases rapidly with increasing density. Maser emission, on the other hand, originates from very small regions under special physical conditions, probably not representative of the bulk of molecular cloud material. In this respect, CN is probably the best candidate to measure field strengths in star forming cores with densities in the range $n \sim 10^5$–$10^6\,\mathrm{cm}^{-3}$.

Field strengths measured by Zeeman effect in HI and OH are of the order of a few times 10 μG up to $\sim 10^2$ μG (see e.g. Bourke et al., 2001, Crutcher et al., 2003), whereas CN observations indicate larger field intensities, up to 0.4–0.7 mG (Crutcher et al. 1999). In OH and H_2O masers the field is much stronger, about 10 mG and 10^2 mG, respectively (Fiebig and Güsten, 1989; Güsten et al., 1994). For comparison, the intensity of the local Galactic magnetic field is ~ 5 μG (Troland and Heiles, 1986).

Polarization of thermal dust emission, on the other hand, provides information about the direction of the component of the magnetic field in the plane of the sky. Thus, mapping the polarization at (sub)millimetre wavelengths is the most reliable mean of probing the magnetic field geometry in molecular clouds. Polarization maps of molecular clouds and filaments associated with regions of star formation have revealed rather uniform patterns of magnetic field-lines (see e.g. Ward-Thompson et al., 2000; Matthews and Wilson, 2002; Crutcher et al., 2003), indicating the presence of a dominating large scale component of the field.

Finally, it has been proposed that a combination of continuum polarimetry with spectroscopic measurements of the ratio of ion and neutral molecular line widths could in principle provide information on the three-dimensional structure of the magnetic field in the weakly ionized parts of molecular clouds (Houde et al., 2002). This method has recently been successfully used to infer the orientation of the magnetic field in the Orion A star forming region (Houde et al., 2004).

Taken together, these observations imply that in the dense condensations within molecular clouds (the so-called cloud cores), the gravitational, thermal, and magnetic energy are roughly of the same order of magnitude (Myers and Goodman, 1988). The magnetic field is therefore expected to play an important role in the evolution of local density enhancements and in their eventual collapse. In this review, we shall focus attention on some aspects of this process, namely: (i) the stability of molecular clouds, (ii) the formation and evolution of condensations within them, (iii) the anisotropic collapse of magnetized unstable cores, and (iv) the substantial braking of the rotation of clouds caused by the magnetic field.

2. Physical Properties of Magnetized Clouds

The high densities measured in dense molecular cores ($n \sim 10^5$ cm^{-3}) are close to the value of the critical density for gravitational (dynamical) collapse according to the classical Jeans criterion. However, large complexes of cores, or entire molecular clouds, having masses much in excess of their associated Jeans masses, cannot be supported by thermal pressure alone. In fact, theoretical considerations (Mestel and Spitzer, 1956; Strittmatter, 1966; Mouschovias, 1976a,b; see the review by McKee et al., 1993) show that a magnetic field of strength B can support a molecular cloud of radius R provided its mass M is less than a magnetic critical mass

$$M_{\rm cr} \simeq 0.12\frac{\Phi}{\sqrt{G}} \simeq 10^3\left(\frac{B}{30\,\mu{\rm G}}\right)\left(\frac{R}{2\,{\rm pc}}\right)^2 M_\odot, \tag{1}$$

where $\Phi = \pi R^2 B$ is the magnetic flux through the cloud. Supercritical clouds ($M > M_{\rm cr}$) cannot be supported by magnetic fields alone, even if they were perfectly frozen in the gas (see Section 3), and they would collapse on a magnetically diluted free-fall timescale. Subcritical clouds ($M < M_{\rm cr}$) are supported by magnetic field against collapse (even if the external pressure increases) and evolve on the time scale that characterizes the diffusion of the magnetic field. On the scale of molecular cloud cores, ($R \sim 0.1$ pc), the magnetic critical mass is of the order of a few $M_\odot$, close to the actual mass of low-mass (Taurus) cores. Therefore, the initial conditions for star formation are characterized by a condition of near magnetic criticality ($M \sim M_{\rm cr}$). Any theory of star formation should then elucidate in the first place how the small fraction of mass of a molecular cloud that is eventually converted into stars (less than ~10%, according to Lada and Lada, 2003) reached this critical state.

In addition to magnetic fields, the turbulent motions inferred by the width of molecular lines are also expected to represent a fierce opposition to gravity in preventing molecular clouds from collapse as a whole on the dynamical time scale

$$t_{\rm dyn} \simeq \frac{1}{\sqrt{G\rho}} \simeq 10^5\left(\frac{n}{10^4\,{\rm cm}^{-3}}\right)^{-1/2} {\rm yr}, \tag{2}$$

as the Jeans criterion would imply. However, both analytical studies and numerical simulations agree that hydro- or magneto-hydrodynamical turbulence in the interstellar medium should decay in about one dynamical time (Gammie and Ostriker 1996; MacLow et al., 1998), therefore requiring a continuous injection of turbulent energy to sustain the clouds over their lifetimes. Magnetic fields, on the other hand, are long lived: in a lightly ionized medium such as a the gas of a molecular cloud (ionization fraction $\sim 10^{-8}$–10^{-7}), ohmic losses are too weak to significantly reduce the magnetic field strength over the cloud's lifetime. For example, for an ionization fraction $\sim 10^{-8}$ and density $\sim 10^5$ cm^{-3}, the ohmic dissipation time of a magnetic field extending on a length scale of ~0.1 pc is of the order of 10^{15} yr.

3. Quasistatic Evolution

Even at the very low levels of ionization characterizing molecular clouds, electrons and ions collide more frequently with each other rather than with much more abundant neutral particles. Collisions with neutrals start to dominate only at densities larger than $\sim 10^5$ cm^{-3} (for ions) and $\sim 10^8$ cm^{-3} (for electrons). At lower densities, and for $T \simeq 10$ K, the gyrofrequency $\omega = eB/mc$ (in CGS units) of a particle of mass m and charge e (either an electron or a molecular ion) is much larger than the collision time τ of this particle with other species. Therefore, the product $\omega\tau$ is usually a large number, both for electrons and ions:

$$\omega_e \tau_e \simeq 10^5 \left(\frac{B}{10\,\mu\mathrm{G}}\right)\left(\frac{n}{10^4\,\mathrm{cm}^{-3}}\right)^{-1/2}, \tag{3}$$

$$\omega_i \tau_i \simeq 10^2 \left(\frac{B}{10\,\mu\mathrm{G}}\right)\left(\frac{n}{10^4\,\mathrm{cm}^{-3}}\right)^{-1/2}. \tag{4}$$

Physically, the condition $\omega\tau \gg 1$ for electrons and ions means that the magnetic field is carried out by the charged particles (*field freezing* in the plasma). Charged grains, due to their larger masses, are much less strongly coupled to the field, even at low densities. However, because of their large collisional cross section (of the order of the geometric cross section), grains contribute significantly to the total frictional force exerted by the charged species on the neutrals at densities larger than $\sim 10^5\,\mathrm{cm}^{-3}$ (Nakano et al., 2002).

Under these physical conditions, the relevant mechanism of magnetic diffusion is the so-called *plasma drift* or *ambipolar diffusion*, originally proposed by Mestel and Spitzer (1956). In this process, the fluid of charged particles with its frozen-in magnetic field can slowly drift with respect to the fluid of neutral particles, the two fluids being collisionally coupled. If $M \simeq M_{\mathrm{cr}}$, the time scale of magnetic flux loss t_{AD} in a cloud where thermal pressure and magnetic forces nearly balance the cloud's self-gravity, is

$$t_{\mathrm{AD}} \simeq \frac{n_i \langle \sigma v \rangle}{\pi G \rho} \simeq 10^7 \left(\frac{x_i}{10^{-7}}\right) \mathrm{yr}, \tag{5}$$

where n_i is the ion density, $\langle \sigma v \rangle$ is the relevant collision rate coefficient, ρ is the cloud's mass density, and x_i is the ionization fraction. In the range $10^3 \lesssim n \lesssim 10^7\,\mathrm{cm}^{-3}$, with $M \simeq M_{\mathrm{cr}}$, this ambipolar diffusion time scale is of the order of several times 10^6 yr, or about one–two orders of magnitude larger than the free-fall time scale.

The effects of the gas turbulence on this process have been considered at various levels, starting with the semi-empirical treatment by Lizano and Shu (1989). In general, the result is that the ambipolar diffusion rate can be enhanced by turbulence because it creates small scale structures which diffuse faster than the original large scale structures (Zweibel, 2002; Fatuzzo and Adams, 2002; Kim and Diamond, 2002; Heitsch et al., 2004). The actual evolution of a magnetized interstellar cloud is made complex by the simultaneous presence of several species, including neutral molecules and atoms (mostly H_2 and He), electrons, molecular and atomic ions, neutral and charged dust grains (Umebayashi and Nakano, 1990). Their relative abundances depend on the ambient conditions, like the total density, the cosmic-ray flux and the ultraviolet radiation field.

A number of authors have performed detailed ambipolar diffusion computations to follow the phase of formation of molecular cloud cores starting from subcritical, low-density initial states (e.g. Nakano, 1976; Lizano and Shu, 1987; Ciolek and Mouschovias, 1999). These studies have shown that molecular cores evolve quasi-statically in a time of the order of $t_{\rm AD}$ towards a magnetically supercritical state in the central region and subcritical in the magnetically supported envelope, characterized by the density profile of the singular isothermal sphere ($n \propto r^{-2}$) modified by a shape function accounting for the anisotropy of magnetic support. These asymptotic configurations, sometimes called a "pivotal state," separate the nearly quasi-static phase of core evolution from the fully dynamic phase of protostellar accretion, that begins when the supercritical central region undergoes dynamical collapse over a time scale of the order of the free-fall collapse and a growing point mass is formed at the core's centre.

4. Dynamical Collapse

The accumulation of matter into denser regions driven by gravity during the ambipolar diffusion phase results in the formation of a flattened and centrally peaked core permeated by a hourglass-shaped magnetic field (see e.g. Fiedler and Mouschovias, 1992, 1993). An important effect of the field on the subsequent collapse dynamics is the strong non-radial Lorentz force resulting from the field configuration, that deflects infalling fluid elements towards the equatorial plane forming an inner non-equilibrium structure ("pseudo-disk") around the central protostar (Galli and Shu, 1993a,b). Recently Allen et al. (2003) performed numerical collapse calculations of the pivotal states calculated by Li and Shu (1996) employing the Zeus-2D magneto-hydrodynamical package (Stone and Norman, 1992). In qualitative agreement with the previous results of Galli and Shu (1993a,b), Allen et al. (2003) found that the infall proceeds at a constant rate and magnetically supported, high density pseudo-disks form in the direction perpendicular to the initial magnetic field.

The collapse of the strongly peaked "pivotal" core proceeds inside-out, with the central regions approaching a $r^{-3/2}$ density profile whereas the envelope maintains the r^{-2} profile produced during the ambipolar diffusion phase. Mass conservation (nr^3 = constant) and strict flux freezing (Br^2 = constant), valid if the collapse takes place on a time scale shorter than the ambipolar diffusion time scale, imply that the field increases as $B \propto n^{2/3}$ during the core collapse, but this scaling would lead to stellar magnetic field strength much in excess of those observed. This fact constitutes a fundamental problem for any theory of star formation: a $\sim 1\ M_\odot$ of interstellar material must reduce its magnetic flux by about 3–5 orders of magnitude to become a magnetic star, or by about eight orders of magnitude to become an ordinary star like the Sun. This *magnetic flux problem* was already recognized by Mestel and Spitzer (1956).

There is little doubt that the bulk of the magnetic field threading the core must have been lost at some stage. In fact, at very high densities ($\sim 10^{11}$ cm^{-3}), theoretical models predict the existence of a *decoupling* phase, where the condition of good coupling of the field in the plasma (flux freezing) breaks down and the neutral fluid ceases to be affected by the magnetic field (through collisions with charged particles) and vice-versa (Nakano and Tademaru, 1972; Nakano et al., 2002). The recent numerical simulations of Desch and Mouschovias (2001) follow the evolution of a low-mass core driven by ambipolar diffusion up to the point when magnetic decoupling starts to occur, and find that a substantial field dissipation takes places at somewhat lower densities, $n \sim 10^{10}$ cm^{-3}. At these high densities, it is possible that the magnetic field lines, strongly pinched by the inward pull of the infalling gas, suffer considerable detachment and reconnection, destroying completely the original field topology, a process first analyzed in this context by Mestel and Strittmatter (1967).

5. Rotation and Magnetic Braking

Besides magnetic fields, any realistic calculation of the collapse of a cloud should include the effects of rotation. In fact, small but detectable levels of rotation have been measured in the outer parts of molecular cloud cores (see e.g. Goodman et al., 1993, Caselli et al., 2002), with typical values corresponding to a ratio of rotational to gravitational energy of $\sim$0.02. Although these velocity gradients are too small to contribute significantly to the support of cloud cores, the dynamical importance of rotation is expected to increase during the collapse of the core, leading ultimately to the formation of a centrifugally supported circumstellar disk around the accreting protostar.

If the angular momentum of a cloud core were strictly conserved during the collapse phase, accretion would be stopped by the centrifugal barrier at a density much lower than the stellar density. In fact, a typical $\sim 1\ M_\odot$ cloud core has about three orders of magnitude more angular momentum than the Solar System, and about one order of magnitude more angular momentum than a wide binary of comparable mass with period $\sim 10^2$ yr (the well-known *angular momentum problem*). In addition, the observed rotational velocities are much lower than expected if the core is condensed out of a low-density background medium with angular momentum conserved.

An efficient mechanism for removing the angular momentum from a cloud is via transport along the magnetic field lines anchored in the ambient medium. If the direction of the magnetic field is parallel to the core's rotation axis, this *magnetic braking* takes place on a time scale

$$t_{\rm braking} \simeq \frac{\Sigma}{2\rho_0 v_{\rm A}} \simeq 10^5 \left(\frac{M}{M_{\rm cr}}\right)\left(\frac{n_0}{10^3\ {\rm cm}^{-3}}\right)^{-1/2} \ {\rm yr}, \tag{6}$$

where Σ is the column density of the cloud, ρ_0 (n_0) and $v_{\rm A}$ are the density and Alfvèn velocity in the ambient medium (Ebert et al. 1960; Mouschovias and Paleologou

1980). Physically, t_{braking} is the time taken by a torsional Alfvèn wave to cross a region with a momentum of inertia equal to that of the core. Since $t_{\text{braking}} < t_{\text{AD}}$, during the quasi-static phase of cloud evolution driven by ambipolar diffusion magnetic braking is in principle able to enforce rotation of the core at the same angular velocity Ω of the envelope (Basu and Mouschovias, 1994). On the other hand, in the supercritical inner parts of the core, the collapse proceeds unimpeded by the magnetic field, and infalling fluid elements are expected to conserve their specific angular momentum J/M.

Particularly suggestive in this respect are the observations collected and analyzed by Ohashi et al. (1997), showing a qualitative agreement with the expected $\Omega =$ constant and $J/M =$ constant trends for cloud cores and infalling envelopes around protostars, respectively, the separation between the two regimes occurring roughly at a spatial scale ~0.03 pc. This value is quite close to the radius of the supercritical core for a cloud of radius 0.75 pc predicted by theoretical models of cloud evolution via ambipolar diffusion (~0.05 pc, see e.g. Desch and Mouschovias, 2001). However, the observed constant value of J/M for infalling envelopes is comparable to the specific angular momentum of a wide binary system, but still about two orders of magnitude larger than the value of J/M of the Solar System. It is clear that additional removal of angular momentum is needed during or after the collapse phase, in order to form a single star. A magneto-centrifugally driven wind from a star or a circumstellar disk is the best candidate for performing this task (Köenigl, 1991).

6. Conclusions

Over the last 20–40 years, observations and theory have contributed to define the importance of the magnetic field at every stage of the process of star formation. In this review, we have briefly discussed some physical processes in which magnetic fields are expected to play a major role in the evolution of interstellar clouds. Let us at least mention in passing other important aspects not considered in this review, like the extra support provided by Alfvèn waves propagating along field lines in a weakly ionized gas (McKee and Zweibel, 1995), the role played by magnetic fields in the process of fragmentation (Mestel, 1985), and in the powering of protostellar winds and outflows (Shu et al., 1988). Some of the ideas discussed here have been incorporated in a coherent theory of "magnetically controlled" star formation, elaborated, with slightly different points of view, e.g. by Mestel (1965, 1985), Shu et al. (1987), and Mouschovias (1991).

The intrinsic difficulty of performing Zeeman measurements of generally weak field strengths, and the high sensitivity required to detect the low levels of polarization of the dust emission, represent the current limitations to a more complete understanding of the spatial distribution of magnetic fields in star forming regions. These limitations in turn make it difficult to assess the validity of the assumptions and to test the accuracy of the results of any theoretical model proposed so far. Future instruments like ALMA will certainly provide the key to answer the remaining open questions.

References

Allen, A., Shu, F.H. and Li, Z.-Y.: 2003, *ApJ* **599**, 351.
Basu, S. and Mouschovias, T.Ch.: 1994, *ApJ* **432**, 720.
Bourke, T.L., Myers, P.C., Robinson, G. and Hyland, A.R.: 2001, *ApJ* **554**, 916.
Caselli, P., Benson, P., Myers, P.C. and Tafalla, M.: 2002, *ApJ* **572**, 238.
Crutcher, R.M., Troland, T.H., Lazareff, B., et al.: 1999, *ApJ* **514**, L121.
Crutcher, R.M., Nutter, D.J., Ward-Thompson, D. and Kirk, J.M.: 2003, *ApJ* **600**, 279.
Desch, S.J. and Mouschovias, T.Ch.: 2001, *ApJ* **550**, 314.
Ebert, R., von Hoerner, S. and Temesvary, S.: 1960, Die Entstehung von Sternen durch Kondensation diffuser Materie, Springer-Verlag, Berlin, p. 315.
Fatuzzo, M. and Adams, F.C.: 2002, *ApJ* **570**, 210.
Fiedler, R.A. and Mouschovias, T.Ch.: 1992, *ApJ* **391**, 199.
Fiedler, R.A. and Mouschovias, T.Ch.: 1993, *ApJ* **415**, 680.
Fiebig, D. and Güsten, R.: 1989, *A&A* **214**, 333.
Galli, D. and Shu, F.H.: 1993a, *ApJ* **417**, 220.
Galli, D. and Shu, F.H.: 1993b, *ApJ* **417**, 243.
Gammie, C.F. and Ostriker, E.C.: 1996, *ApJ* **466**, 814.
Goodman, A.A., Benson, P.J., Fuller, G.A. and Myers, P.C.: 1993, *ApJ* **406**, 528.
Güsten, R., Fiebig, D. and Uchida, K.I.: 1994, *A&A* **286**, L51.
Heitsch, F., Zweibel, E.G., Slyz, A.D. and Devriendt, J.E.G.: 2004, *ApJ* **603**, 165.
Houde, M., Bastien, P., Dotson, J.L., et al.: 2002, *ApJ* **569**, 803.
Houde, M., Dowell, C.D., Hildebrand, R.H., et al.: 2004, *ApJ* **604**, 717.
Kim, E.-J. and Diamond, P.H.: 2002, *ApJ* **578**, 113.
Königl, A.: 1991, *ApJ* **370**, L39.
Lada, C.J. and Lada, E.A.: 2003, *ARAA* **41**, 57.
Li, Z.-Y. and Shu, F.H.: 1996, *ApJ* **472**, 211.
Lizano, S. and Shu, F.H.: 1989, *ApJ* **342**, 834.
MacLow, M.-M., Klessen, R.S., Burkert, A., et al.: 1998, *Phys. Rev. Lett.* **80**, 2754.
Matthews, B.C. and Wilson, C.D.: 2002, *ApJ* **574**, 822.
McKee, C.F., Zweibel, E.G., Heiles, C. and Goodman, A.A.: 1993, in: E. Levy and J. Lunine (eds.), *Protostars and Planets III*, The University of Arizona Press, Tucson, p. 327.
McKee, C.F. and Zweibel, E.G.: 1995, *ApJ* **440**, 686.
Mestel, L. and Spitzer, L.: 1956, *MNRAS* **116**, 505.
Mestel, L.: 1965, *QJRAS* **6**, 265.
Mestel, L. and Strittmatter, P.A.: 1967, *MNRAS* **137**, 95.
Mestel, L.: 1985, in: D.C. Black and M.S. Mattews (eds.), *Protostars and Planets II*, The University of Arizona Press, Tucson, p. 320.
Mouschovias, T.Ch.: 1976a, *ApJ* **206**, 753.
Mouschovias, T.Ch.: 1976b, *ApJ* **207**, 141.
Mouschovias, T.Ch. and Paleologou, E.V.: 1980, *ApJ* **237**, 877.
Mouschovias, T.Ch.: 1996, in: K. Tsinganos (ed.), *Solar and Astrophysical Magnetohydrodynamic Flows*, Kluwer, Dordrecht, p. 475.
Mouschovias, T.Ch.: 1991, in: C.J. Lada and N.D. Kylafis (eds.), *The Physics of Star Formation and Early Stellar Evolution*, Kluwer, Dordrecht, p. 61, 449.

Myers, P.C. and Goodman, A.A.: 1988, *ApJ* **329**, 392.
Nakano, T.: 1976, *PASJ* **28**, 355.
Nakano, T. and Tademaru, E.: 1972, *ApJ* **173**, 87.
Nakano, T., Nishi, R. and Umebayashi, T.: 2002, *ApJ* **573**, 199.
Ohashi, N., Hayashi, M., Ho, P.T.P., et al.: 1997, *ApJ* **488**, 317.
Shu, F.H., Adams, F.C. and Lizano, S.: 1987, *ARAA* **24**, 23.
Shu, F.H., Lizano, S., Ruden, S.P. and Najita, J.: 1988, *ApJ* **328**, L19.
Stone, J.M. and Norman, M.L.: 1992, *ApJS* **80**, 75.
Strittmatter, P.A.: 1966, *MNRAS* **132**, 359.
Troland, T.H. and Heiles, C.: 1986, *ApJ* **301**, 339.
Umebayashi, T. and Nakano, T.: 1990, *MNRAS* **243**, 103.
Ward-Thompson, D., Kirk, J.M., Crutcher, R.M., et al.: 2000, *ApJ* **537**, L135.
Zweibel, E.G.: 2002, *ApJ* **567**, 962.

VLBI WATER MASER PROPER MOTION MEASUREMENTS IN STAR-FORMING REGIONS

J.M. TORRELLES[1], N. PATEL[2], J.F. GÓMEZ[3], G. ANGLADA[4] and L. USCANGA[5]

[1]*Instituto de Ciencias del Espacio (CSIC)-IEEC, C/ Gran Capità 2-4, Barcelona, Spain; E-mail: torrelles@ieec.fcr.es*

[2]*Harvard-Smithsonian Center for Astrophysics, 60 Garden Street, Cambridge, MA, USA*

[3]*Laboratorio de Astrofísica Espacial y Física Fundamental (INTA), Apartado de Correos 50727, Madrid, Spain*

[4]*Instituto de Astrofísica de Andalucía (CSIC), Apartado de Correos 3004, Granada, Spain*

[5]*Instituto de Astronomía (UNAM), Apdo Postal 70-264, Cd. Universitaria, Mexico DF, Mexico*

(Received 16 April 2004; accepted 15 June 2004)

Abstract. We review some of the recent water maser proper motion measurements in star-forming regions performed through VLBI multi-epoch observations. These observations are starting to reveal exciting perspectives, providing the full kinematics of the gas within the outflows/circumstellar disks around YSOs at scales of AUs, discovering new phenomena (e.g., isotropic mass ejections, water maser "micro-structures" exhibiting remarkable coherent and well ordered spatio-kinematical behavior at AU scale), and opening new, puzzling questions related to the early stellar evolution.

Keywords: ISM: jets and outflows, masers–stars: formation

1. Introduction

We know that young stellar objects (YSOs) are characterized by powerful very collimated jets, Herbig-Haro (HH) emission, molecular outflows, maser emission (e.g., H_2O, OH, CH_3OH), and radio continuum emission (e.g., Anglada, 1995, 1996, and contributions by Brand, Cesaroni, Cohen, Gray, Hoare, Niezurawska, Pestalozzi, and Sobolev in this workshop). All these phenomena can be explained within a general scenario of an accretion disk surrounding a protostar, ejecting collimated outflows through the poles of the disk and its subsequent interaction with the ambient medium. This scenario seems to be valid, as a first approach, from low to intermediate and high-mass stars (e.g., Shu et al., 1987; Garay and Lizano, 1999; Osorio et al., 1999, and Cesaroni in this workshop). However, our further knowledge of the earliest stages of stellar evolution has been hampered by the lack of data at the scales of a few AUs, where relevant physical phenomena are expected to occur (e.g., accreting processes, rotating motions, outflow collimation).

These observational limitations provide the main motivation to observe water masers ($6_{16} \rightarrow 5_{23}$; $\nu = 22235.080$ MHz) toward YSOs with extremely high angular resolution. In fact, the ambient medium is transparent to the H_2O maser emission at 1.3 cm, and such emission is compact and strong enough (with brightness temperatures exceeding 10^{10} K), that can be observed with very high angular resolution ($\simeq$0.4 mas) using very long baseline interferometry (VLBI) techniques (e.g., Reid and Moran, 1981). In addition, water masers are good tracers of warm ($\sim$500 K) and dense ($\sim 10^8$–10^9 cm^{-3}) molecular gas (Elitzur et al., 1992; Claussen, 2002; Chandler, 2003). These physical conditions of temperature and density are reached in the shocked gas compressed by winds (Chernin, 1995; Torrelles et al., 1997; Furuya et al., 2000; Moscadelli et al., 2003), in the hot dense infalling gas after the accretion shock (Menten and van der Tak, 2004), as in circumstellar disks (Fiebig et al., 1996; Torrelles et al., 1996, 1998; Seth et al., 2002; Brand et al., 2003). Interestingly, these physical conditions are also reached in galaxies, where water masers trace both outflows and accretion disks (e.g., Miyoshi et al., 1995; Claussen et al., 1998a; Greenhill et al., 2003, and Tarchi and Henkel in this workshop).

In all these scenarios of masers tracing outflowing gas and/or rotating/infalling gas motions in circumstellar disks around YSOs, proper motions are expected. For example, masers moving with velocities $\geq$10 km s^{-1} will result in proper motions $\geq$0.4 mas/month for a source at a distance of 500 pc. These proper motions can be measured through VLBI multi-epoch observations. In general, proper motion measurements of H_2O masers are very limited by strong time variability of their intensity (e.g., Brand et al., 2003); however, the very high angular resolution of the VLBI observations allows to use short time intervals of a few weeks, therefore minimizing the time-dependent effects.

In this paper we review recent H_2O maser proper motion measurements toward six star-forming regions, from low to intermediate and high-mass stars (IRAS 05413-014, S106 FIR, NGC2071 IRS3, IRAS 21391 + 5802, Cepheus A, and W75N). Other very important results and excellent reviews on water maser proper motion measurements can be found in Imai et al. (2000, 2002), Claussen (2002), and Chandler (2003). All these observations have shared some common general goals, as to observe as close as possible the central engine responsible for the phenomena associated with the star formation, to determine the full motions of the outflow/disk at AU scales, and to discover new phenomena related to the excitation of water masers.

2. Water Maser Proper Motion Measurements

2.1. IRAS 05413–014

This is a young low-mass star (probably a class 0 protostar) with a highly collimated jet. The jet is observed in H_2 at infrared wavelengths (2.12 μm) up to scales of $\simeq 4'$ ($\simeq$100,000 AU), and emerges from a circumstellar disk observed in ammonia lines (Figure 1a; Zinnecker et al., 1998; Wiseman et al., 2001).

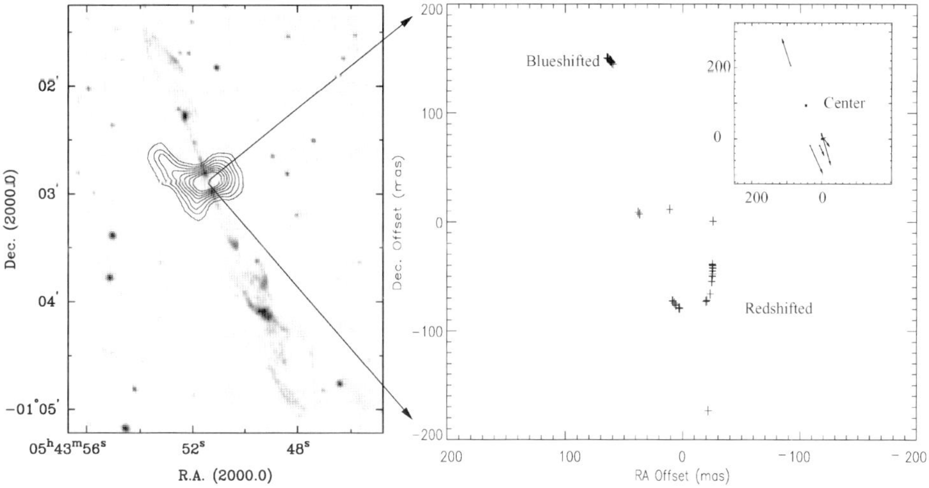

(a) (b)

Figure 1. (a) (*left*): H_2 image (2.12 μm) of the HH 212 jet (grey scale; Zinnecker et al., 1998) in IRAS 05413−014, with the ammonia map of the circumstellar disk (contours; Wiseman et al., 2001) overlapped; (b) (*right*): Close-up of the central region showing the water maser positions and proper motions (inset in the top right-corner of the figure) measured with the VLBA. The dot marked "Center" in the inset is located about the offset position (+35, +25) on the large map (figure adapted from Claussen et al., 1998b).

Water maser emission has been mapped with the very long baseline array (VLBA) of the National Radio Astronomy Observatory (NRAO) by Claussen et al. (1998b). These authors, through multi-epoch observations spread over a time span of two months, found in the innermost region a jet-like water maser bipolar structure of $\simeq$300 mas ($\simeq$120 AU) in extent, with blueshifted radial velocities toward the northeast and redshifted velocities toward the southwest. Spatial velocities are $\simeq$60 km s^{-1}, as derived from the proper motion measurements (Figure 1b). The jet traced by the water masers is almost on the plane of the sky ($i \geq 85°$), as derived from the ratio of the proper motion velocity to the line-of-sight velocity of the masers with respect to the cloud velocity. These VLBA observations indicate that the water masers are tracing the base of the large-scale outflow observed in H_2, implying that jet formation, acceleration, and collimation take place within $\simeq$50 AU radius of the central YSO, which is an important input for theoretical models trying to explain the generation and collimation of the outflows (Claussen et al., 1998b).

2.2. S106 FIR

S106 FIR is a low/intermediate mass star (probably a class 0 protostar) with an estimated total luminosity in the range of $\simeq$ 24–1000 $L_\odot$ (Richer et al., 1993;

IRS 3. Further VLBA multi-epoch water maser observations could help to clarify this important issue.

2.4. IRAS 21391+5802

This is an intermediate mass star which has been studied through VLBA multi-epoch water maser observations by Patel et al. (2000), showing a bipolar outflow extending over $\simeq 0\farcs7$ ($\simeq$500 AU) with spatial velocities of $\simeq$40 km s^{-1} and tracing the base of a more extended molecular outflow observed up to scales of 50,000 AU by Beltrán et al. (2002). One of the most interesting results of these VLBA observations is the detection of a circular loop of masers of 1 AU radius, interpreted by Patel and collaborators as a wind-shell. This would imply that the outflow is poorly collimated at distances close (a few AUs) to the central object, but it is highly collimated at larger distances (few hundred AUs). Furthermore, these authors find that water masers that are more distant from the central source move faster than the closer ones, implying an acceleration of the gas. All these results provide important observational constraints for the generation and collimation mechanism of the outflow.

2.5. Cepheus A

This is a high-mass star forming region exhibiting many of the phenomena related with YSOs (Gómez et al., 1999). At the center of a high-density molecular core there are multiple radio continuum sources and strong H_2O and OH masers (Hughes and Wouterloot, 1984; Rodríguez et al., 1994; Hughes et al., 1995; Garay et al., 1996; Torrelles et al., 1996, and contribution by Niezurawska in this workshop). HW2, which is the brightest radio continuum source of the region, is a thermal radio jet (Rodríguez et al., 1994) associated with a very bright cluster of water masers observed with the VLA (Torrelles et al., 1996). These masers are distributed in a band perpendicular to the radio jet. Both the spatial distribution of the masers and the velocity gradients observed along the major and minor axes of the band suggested that they are tracing a rotating and contracting circumstellar disk around the central YSO.

These water masers were observed with the VLBA in three epochs covering a time interval of two months (Torrelles et al., 2001a,b). Although the spatial distribution and radial velocities of the masers as seen with the two sets of VLBA and VLA data coincide quite well, the failure of the VLBA observations to detect those masers located to the east of HW2 made it difficult for us to test the scenario of a water maser disk around this YSO. However, the VLBA observations revealed unexpected phenomena, in particular the discovery that some of the single water masers detected previously with the VLA unfold into remarkable linear/arcuate microstructures with proper motions. The most important result of these VLBA observations has been the detection of a bright arc-like structure of $\simeq$100 mas size ($\simeq$72 AU) located $\simeq$0$''$.7 to the south of HW2, and persisting in the three epochs. This arc of masers is extremely well fitted by a circle of 62 AU radius to an accuracy of one part in a thousand. Moreover, this arc structure is constituted by a string of smaller linear structures of maser spots with sizes $\simeq$ 0.4–1 AU ("building-blocks") and different orientations, all of which are tangential to the arc curvature (Figure 2). Proper motions of the water masers indicate a uniform expansion of 9 km s^{-1} perpendicular to the arc. Both the high degree of symmetry of the arc and its small dynamical age (33 years) suggest that it represents the limb brightened parts of a singular, short lived, and episodic spherical ejection (explosive event) driven by a YSO. This protostar would be located at the center of the circle which fits the arc structure, with the smaller linear "building-blocks" being most likely highly flattened surfaces defining shock fronts (Torrelles et al., 2001a,b). In fact, a weak 3.6 cm continuum source (0.2 mJy) has been detected with the VLA nearly coinciding with the center of the circle and proposed to be the powering source of the expanding bubble of masers, although its nature is still unknown (Curiel et al., 2002).

The additional physical relevance of these VLBA results is found in the fact that isotropic ejections are difficult to explain within the current paradigm of star formation, where we expect to observe collimated outflows due to the presence of circumstellar disks. At the moment, the origin of this spherical expanding bubble remains unknown, but it could represent new constraints for theories on how stars evolve in their early stages.

2.6. W75N

This is also a high-mass star forming region with a large-scale molecular outflow (seen in CO) extending over several parsecs. At the center of the molecular outflow there are several millimeter continuum sources, ultracompact HII regions, and strong H_2O and OH maser emission (Haschick et al., 1981; Fischer et al., 1985; Hunter et al., 1994; Torrelles et al., 1997; Davis et al., 1998; Hutawarakorn et al., 2002; Slysh et al., 2002; Shepherd et al., 2003; Shepherd et al., 2004, Niezurawska in this workshop). In particular, at the much smaller scales of arcseconds (few thousand of AUs), there are three radio continuum sources, VLA 1, VLA 2, and VLA 3, which are probably excited by early B-type stars, and two clusters of water masers, one associated with VLA 1, and the other one with VLA 2 (Torrelles et al., 1997). The masers associated with VLA 1 are distributed in a cluster of $\simeq$1$''$ in size ($\simeq$2000 AU), extending along the major axis of the observed elongated radio continuum emission, suggesting that VLA 1 could be a radio jet in which the water masers are tracing the base of a collimated outflow. On the contrary, in the case of VLA 2, the associated water masers are very tightly grouped in space forming a shell of $\simeq$0$''$.2 size ($\simeq$400 AU) around the compact radio continuum emission. Given the compactness of the water maser distribution around VLA 2 it was tentatively suggested that these masers could trace bound motions in a circumstellar disk (Torrelles et al., 1997).

For testing the scenarios proposed for the water maser distribution around VLA 1 (outflow) and VLA 2 (disk), VLBA multi-epoch water maser observations were carried out by Torrelles et al. (2003), revealing proper motions of ninety water masers. The most remarkable result of these VLBA observations is that while the water maser proper motions in VLA 1 are parallel to the elongated radio continuum emission, in VLA 2 the masers are distributed in a shell moving outwards in multiple directions from the central radio continuum source. With more detail, the VLBA water masers around VLA 1 form a linear structure of $\simeq$2000 AU with a small spatial dispersion in width of 200 AU. The measured proper motions parallel to the elongated radio continuum emission are $\simeq$19 km s^{-1}. This indicates that in VLA 1 there is a highly collimated outflow, supporting the previous interpretation based on the VLA observations. In contrast, in the case of VLA 2, masers are moving outwards from the central continuum source in multiple directions with a mean proper motion vector of $\simeq$28 km s^{-1}. Furthermore, the observed motions of the VLA 2 water masers would require a binding mass of the order of 100 $M_\odot$, which is not observed. Hence, these proper motions are consistent with expansion, discarding that the masers trace bound motions in a circumstellar disk as suggested by the previous VLA observations.

The uniqueness of the VLBA results on this source is the well differentiated outflow geometry found simultaneously in two similar YSOs (early B-type stars), separated by only 1400 AU, suggesting that outflow collimation is not only a consequence of ambient conditions (because both objects share the same molecular environment), but something intrinsic to the individual evolution of stars.

Several scenarios have been considered by Torrelles et al. (2003) for the driving agent of the VLA 2 water maser shell, concluding that the most likely one is a wind-driven shell with a wind mass loss rate $\dot{M}_{\rm w} \simeq 8 \times 10^{-7} M_\odot {\rm yr}^{-1}$, and velocity $V_{\rm w} \simeq 100$ km s^{-1}. These values are consistent with those expected from early B-type stars. However, as in Cepheus A, the open question is *how and when the non-collimated wind ejection stage occurs in the evolution of YSOs*? With regard to this important point, it has been argued (Torrelles et al., 2003) that since the non-collimated motions around VLA 2 are observed at smaller scales in comparison with the scale where the collimated motions are observed around VLA 1, VLA 2 could be younger than VLA 1. In addition, since the spatial dispersion of the water masers from the major axis of the VLA 1 radio jet ($\simeq$200 AU) is similar to the scale where the outflow of VLA 2 is observed, VLA 2 could evolve in the future into a collimated jet as VLA 1. In any case, we think that this is a clear example where observations at (sub) millimeter wavelengths with angular resolutions of $\simeq 0\farcs1$ are crucial for determining the relative stage of evolution of these two YSOs to know whether isotropic ejections are produced before or after the stage of highly collimated outflows.

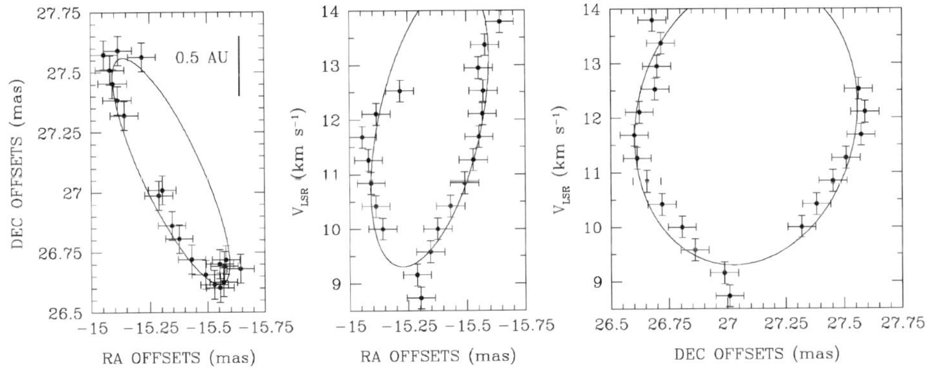

Figure 3. Water maser microstructure observed with the VLBA in the W75N VLA 2 water maser shell. This microstructure (only detected in the second epoch of the VLBA observations, Torrelles et al., 2003) fits into a ring of 1 AU radius, with an expansion velocity of 2.7 km s^{-1}, and seen with an inclination angle of 76° (Uscanga et al., 2004).

Finally, another very important result obtained with the VLBA water maser observations has been the discovery that masers in the VLA 2 shell, at the very small scales of AU, tend also to form very coherent and well-ordered spatio-kinematical "microstructures," as in the case of Cepheus A (see Section 2.5). In Figure 3 we show one of these remarkable "microstructures" which exhibits a ring-like shape. This structure fits extremely well into a ring of 1 AU radius, with an expansion velocity of 2.7 km s^{-1}, and seen with an inclination angle of 76° (Uscanga et al., 2004). We do not know yet the origin of this outstanding spatio-kinematical coherence at these very small scales, but it suggests that this is a fundamental scale for the excitation of water masers. Anglada et al. (1996) found that the most luminous water masers are associated with a larger perturbation of the molecular gas, as traced by CS and ammonia line widths. Furthermore, these authors found a correlation between the observed line widths of both the CS and ammonia lines and the maser luminosity, $L(H_2O) \propto (\Delta V)^3$, that is in agreement with what is expected if the maser luminosity is a consequence of the dissipation in the ambient cloud of the kinetic energy of outflowing maser condensations (Reid and Moran, 1988). Anglada et al. (1996) proposed an scenario in which water maser emission is intrinsically associated with an increase in the thermal and turbulent energy of the ambient cloud. We speculate that perhaps in these VLBA "microstructures" we are already reaching the scale where turbulent motions are resolved.

3. Final Remarks

VLBA multi-epoch water maser observations are a very powerful tool to study the structure and kinematics of the gas very close to the central engine of YSOs.

These observations make it possible to study both the base of large-scale outflows in star-forming regions and the associated circumstellar disks. Most importantly, these high-angular resolution observations have just started to reveal new phenomena not only related with the excitation of water masers but also with the earliest stages of YSOs, opening new puzzling questions, such as: What is the origin of the powerful isotropic ejections? Are isotropic ejections produced before or after the stage of well collimated ejections? What is the origin of the remarkable coherent and well-ordered spatio-kinematical behavior of the water maser emission at AU scales?

Acknowledgements

GA, JFG, and JMT acknowledge support from MCYT grant AYA2002-00376 (including FEDER funds). GA acknowledges support from Junta de Andalucía. LU ackowledges support from CONACyT grant 145174 and DGEP (UNAM) grant 101301.

References

Anglada, G.: 1995, *Rev Mex AASC* **1**, 67.
Anglada, G.: 1996, in: A.R. Taylor and J.M. Paredes (eds.), *Proceedings of the ASP Conference Series on Radio Emission from the Stars and the Sun*, Vol. 93, p. 3.
Anglada, G., Estalella, R., Pastor, J., Rodríguez, L.F. and Haschick, A.D.: 1996, *ApJ* **463**, 205.
Beltrán, M.T., Girart, J.M., Estalella, R., Ho, P.T.P. and Palau, A.: 2002, *ApJ* **573**, 246.
Brand, J., Cesaroni, R., Comoretto, G., Felli, M., Palagi, F., Palla, F. and Valdettaro, R.: 2003, *A&A* **407**, 573.
Chandler, C.J.: 2004 in press, in: J.D. Romney and M.J. Reid (eds.), *Proceedings of the ASP Conference Series on Future Directions in High Resolution Astronomy: The 10th Anniversary of the VLBA*.
Chernin, L.M.: 1995, *ApJ* **440**, L97.
Claussen, M.J.: 2002, in: V. Migenes and M.J. Reid (eds.), *Proceedings of the ASP Conference Series, (IAU Symposium No. 206)* Vol. 206, p. 27.
Claussen, M.J., Diamond, P.J., Braatz, J.A., Wilson, A.S. and Henkel, C.: 1998a, *ApJ* **500**, L129.
Claussen, M.J., Marvel, K.B., Wootten, A. and Wilking, B.A.: 1998b, *ApJ* **507**, L79.
Curiel, S., Trinidad, M.A., Cantó, J., Rodríguez, L.F., Torrelles, J.M., Ho, P.T.P., Patel, N., Greenhill, L., Gómez, J.F., Garay, G., Hernández, L., Contreras, M.E. and Anglada, G.: 2002, *ApJ* **564**, L35.
Davis, C.J., Moriarty-Schieven, G., Eislöffel, J., Hoare, M.G. and Ray, T.P.: 1998, *AJ* **115**, 1118.
Eiroa, C., Elsässer, H. and Lahulla, J.F.: 1979, *A&A* **74**, 89.
Elitzur, M., Hollenbach, D.J. and McKee, C.F.: 1992, *ApJ* **394**, 221.
Fiebig, D., Duschl, W.J., Menten, K.M. and Tscharnuter, W.M.: 1996, *A&A* **310**, 199.
Fischer, J., Sanders, D.B., Simon, M. and Solomon, P.M.: 1985, *ApJ* **293**, 508.
Furuya, R.S., Kitamura, Y., Saito, M., Kawabe, R. and Wootten, H.A.: 1999, *ApJ* **525**, 821.
Furuya, R.S., Kitamura, Y., Wootten, H.A., Claussen, M.J., Saito, M., Marvel, K.B. and Kawabe, R.: 2000, *ApJ* **542**, L135.
Garay, G. and Lizano, S.: 1999, *PASP* **111**, 1049.
Garay, G., Ramírez, S., Rodríguez, L.F., Curiel, S. and Torrelles, J.M.: 1996, *ApJ* **459**, 193.

Gómez, J.F., Sargent, A., Torrelles, J.M., Ho, P.T.P., Rodríguez, L.F., Cantó, J. and Garay, G.: 1999, *ApJ* **514**, 287.
Greenhill, L.J., Booth, R.S., Ellingsen, S.P., Herrnstein, J.R., Jauncey, D.L., McCulloch, P.M., Moran, J.M., Norris, R.P., Reynolds, J.E. and Tzioumis, A.K.: 2003, *ApJ* **590**, 162.
Haschick, A.D., Reid, M.J., Burke, B.F. and Moran, J.M.: 1981, *ApJ* **244**, 76.
Hughes, V.A., Cohen, R.J. and Garrington, S.: 1995, *MNRAS* **272**, 469.
Hughes, V.A. and Wouterloot, J.G.A.: 1984, *ApJ* **276**, 204.
Hunter, T.R., Taylor, G.B., Felli, M. and Tofani, G.: 1994, *A&A* **284**, 215.
Hutawarakorn, B., Cohen, R.J. and Brebner, G.C.: 2002, *MNRAS* **330**, 349.
Imai, H., Kameya, O., Sasao, T., Miyoshi, M., Deguchi, S., Horiuchi, S. and Asaki, Y.: 2000, *ApJ* **538**, 751.
Imai, H., Watanabe, T., Omodaka, T., Nishio, M., Kameya, O., Miyaji, T. and Nakajima, J.: 2002, *PASJ* **54**, 741.
Menten, K.M. and van der Tak, F.F.S.: 2004, *A&A* **414**, 289.
Miyoshi, M., Moran, J., Herrnstein, J., Greenhill, L., Nakai, N., Diamond, P. and Inoue, M.: 1995, *Nature* **373**, 127.
Moscadelli, L., Cesaroni, R. and Rioja, M.J.: 2000, *A&A* **360**, 663.
Osorio, M., Lizano, S. and D'Alessio, P.: 1999, *ApJ* **525**, 808.
Patel, N.A., Greenhill, L.J., Herrnstein, J., Zhang, Q., Moran, J.M., Ho, P.T.P. and Goldsmith, P.F.: 2000, *ApJ* **538**, 268.
Persson, S.E., Geballe, T.R., Simon, T., Lonsdale, C.J. and Baas, F.: 1981, *ApJ* **251**, L85.
Reid, M.J. and Moran, J.M.: 1981, *ARA&A* **19**, 231.
Reid, M.J. and Moran, J.M.: 1988, in: G.L. Verschuur and K. Kellermann (eds.), *Galactic and Extra Galactic Radio Astronomy*, Springer, New York, p. 255.
Richer, J.S., Padman, R., Ward-Thompson, D., Hills, R.E. and Harris, A.I.: 1993, *MNRAS* **262**, 839.
Rodríguez, L.F., Garay, G., Curiel, S., Ramírez, S., Torrelles, J.M., Gómez, Y. and Velázquez, A.: 1994, *ApJ* **430**, L65.
Seth, A.C., Greenhill, L.J. and Holder, B.P.: 2002, *ApJ* **581**, 325.
Shepherd, D.S., Kurtz, S.E. and Testi, L.: 2004, *ApJ* **601**, 952.
Shepherd, D.S., Testi, L. and Stark, D.P.: 2003, *AJ* **584**, 882.
Shu, F.H., Adams, F. and Lizano, S.: 1987, *ARA&A* **25**, 23.
Slysh, V.I., Migenes, V., Val'tts, I.E., Lyubchenko, S.Y., Horiuchi, S., Altunin, V.I., Fomalont, E.B. and Inoue, M.: 2002, *ApJ* **564**, 317.
Torrelles, J.M., Gómez, J.F., Rodríguez, L.F., Curiel, S., Anglada, G. and Ho, P.T.P.: 1998, *ApJ* **505**, 756.
Torrelles, J.M., Gómez, J.F., Rodríguez, L.F., Curiel, S., Ho, P.T.P. and Garay, G.: 1996, *ApJ* **457**, L107.
Torrelles, J.M., Gómez, J.F., Rodríguez, L.F., Ho, P.T.P., Curiel, S. and Vázquez, R.: 1997, *ApJ* **489**, 744.
Torrelles, J.M., Patel, N., Anglada G., Gómez, J.F., Ho, P.T.P., Cantó, J., Curiel, S., Lara, L., Alberdi, A., Garay, G. and Rodríguez, L.F.: 2003, *ApJ* **598**, L115.
Torrelles, J.M., Patel, N.A., Gómez, J.F., Ho, P.T.P., Rodríguez, L.F., Anglada, G., Garay, G., Greenhill, L., Curiel, S. and Cantó, J. 2001a, *Nature*, **411**, 277.
Torrelles, J.M., Patel, N.A., Gómez, J.F., Ho, P.T.P., Rodríguez, L.F., Anglada, G., Garay, G., Greenhill, L., Curiel, S. and Cantó, J.: 2001b, *ApJ* **560**, 853.
Uscanga, L. et al.: 2004, in preparation.
Wiseman, J., Wootten, A., Zinnecker, H. and McCaughrean, M.: 2001, *ApJ* **550**, L87.
Zinnecker, H., McCaughrean, M.J. and Rayner, J.T.: 1998, *Nature* **394**, 862.

HIGH-RESOLUTION OBSERVATIONS IN B1-IRS: AMMONIA, CCS AND WATER MASERS

ITZIAR DE GREGORIO-MONSALVO[1,2], CLAIRE J. CHANDLER[2], JOSÉ F. GÓMEZ[1], THOMAS B.H. KUIPER[3], JOSÉ M. TORRELLES[4] and GUILLEM ANGLADA[5]

[1]*Laboratorio de Astrofísica Espacial y Física Fundamental (INTA), Madrid, Spain; E-mail: itziar@laeff.esa.es*

[2]*National Radio Astronomy Observatory, Socorro, USA*

[3]*Jet Propulsion Laboratory, California Institute of Technology, USA*

[4]*Instituto de Ciencias del Espacio (CSIC) and Institut d'Estudis Espacials de Catalunya, Barcelona, Spain*

[5]*Instituto de Astrofísica de Andalucía (CSIC), Granada, Spain*

(Received 16 April 2004; accepted 15 June 2004)

Abstract. We present a study of the structure and dynamics of the star-forming region B1-IRS (IRAS 03301 + 3057) using the properties of different molecules at high angular resolution (~4″). We have used VLA observations of NH_3, CCS, and H_2O masers at 1 cm. CCS emission shows three clumps around the central source, with a velocity gradient from red to blue-shifted velocities towards the protostar, probably due to the interaction with outflowing material. Water maser emission is elongated in the same direction as a reflection nebula detected at 2 μm by 2MASS, with the maser spots located in a structure of some hundreds of AU from the central source, possibly tracing a jet. We propose a new outflow model to explain all our observations, consisting of a molecular outflow near the plane of the sky. Ammonia emission is extended and anticorrelated with CCS. We have detected for the first time this anticorrelation at small scales (1400 AU) in a star-forming region.

Keywords: stars: formation, pre-main sequence, ISM: individual (B1-IRS), jets and outflows, kinematics and dynamics, molecules

1. Motivation

Molecular spectral line observations are a powerful tool for obtaining information about kinematics, temperature and density in star forming clouds. The combination of the observations of several molecules can be used to give a complete information of the environment of protostars and the stage of their evolution.

CCS molecular lines are very useful for making studies related to the structure and the physical conditions of the clouds, since they are not very opaque but they are intense and abundant in these regions, and its lack of hyperfine structure makes CCS a good molecule for dynamical studies (Saito et al., 1987; Suzuki et al., 1992). CCS is a high-density tracer, like NH_3. The spatial distribution of these two molecules can be used as a clock to date the age of the clouds, since a spatial anticorrelation between

Astrophysics and Space Science **295:** 65–70, 2004.

these two molecules has been observed in dark clouds, and has been interpreted in terms of chemical evolution (Velusamy et al., 1995; Kuiper et al., 1996).

Mass-loss phenomena are common in the earliest stages of stellar evolution. This activity can be traced by water maser emission (Rodríguez et al., 1980). Moreover these masers provide a good characterization of the age of low-mass young stellar objects (YSOs), since class 0 sources are the most probable candidates to harbor this emission (Furuya et al., 2001).

In order to diagnose the physical conditions around YSOs, and study the dynamics and the stage of evolution, we have analyzed the properties of CCS, NH_3, and water masers in the cloud surrounding the far-infrared source B1-IRS (IRAS 03301 + 3057), a class 0 source (Hirano et al., 1997) located in the B1 molecular cloud.

2. Observations

Simultaneous observations of the $J_N = 2_1\text{–}1_0$ transition of CCS (rest frequency = 22,345.388 MHz) and the $6_{16}\text{–}5_{23}$ transition of H_2O (rest frequency = 22,235.080 MHz) were carried out on April 4, 2003 using the very large array (VLA) of the National Radio Astronomy Observatory[1] (USA) in its D configuration.

We have also processed VLA archive data of the NH_3 (1,1) transition (rest frequency = 23,694.496 MHz). The observations were made on August 13, 1988, with the D configuration.

In addition to all these radio data, we have retrieved a K-band image from the image data of the Two Micron All Sky Survey (2MASS), to get a better position for the infrared source. This image was processed with AIPS, and convolved with a Gaussian of 5″ FWHM, to search for extended emission.

3. Results

The most interesting characteristics of the CCS emission are its clumpy spatial distribution and the observed velocity gradient. There are three main clumps surrounding the central source, all of them are red-shifted with respect to the systemic velocity of the B1 core ($V_{LSR} = 6.3$ km s^{-1}, Hirano et al., 1997). The molecular gas in all these clumps shows clear velocity gradients, with less red-shifted velocities at positions closer to the central object (see Figure 1).

The ammonia emission is very extended (>2′) and clumpy. The general tendency observed consists in a spatial anticorrelation with respect to the CCS emission, never reported before at such small scales (∼4″, 1400 AU at 350 pc). We think that this anticorrelation is related to the chemical evolution of the source, but we need to continue with this kind of studies in other sources similar to B1-IRS to derive consistent conclusions about the meaning of this distribution.

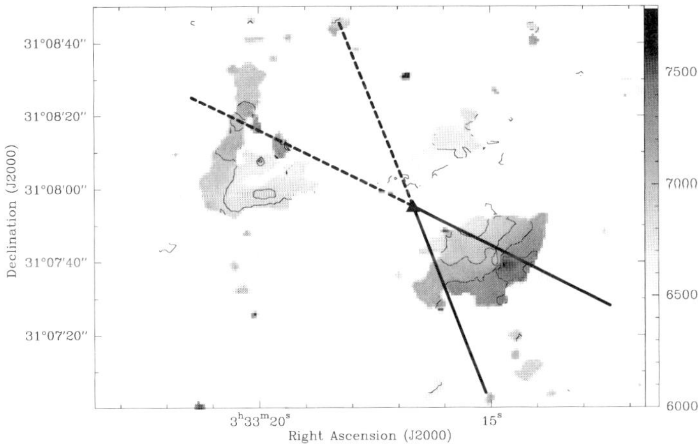

Figure 1. CCS velocity map. Gray scale contours are drawn from 6.0×10^3 to 7.8×10^3 m s^{-1}. Levels are represented from 6.0×10^3 to 7.8×10^3 m s^{-1} by 0.2×10^3 m s^{-1}. The triangle represents the position of the 2 μm point source. The proposed CO outflow configuration is represented by a cone, with solid lines associated with the blue-shifted CO outflow and the dashed lines associated with the red-shifted one, with the lobes almost in the plane of the sky.

Our water maser observation shows a cluster of 23 spots, most of them are red-shifted with respect to the main cloud velocity. They form an elongated structure (see Figure 2), with a length of $\sim$1.″3 (455 AU at 350 pc).

We have detected a point-like source at 2 μm from the 2MASS archive. Convolving the K-band image with a Gaussian of 5″ FWHM, we have noticed an extended emission elongated southwest from the point source.

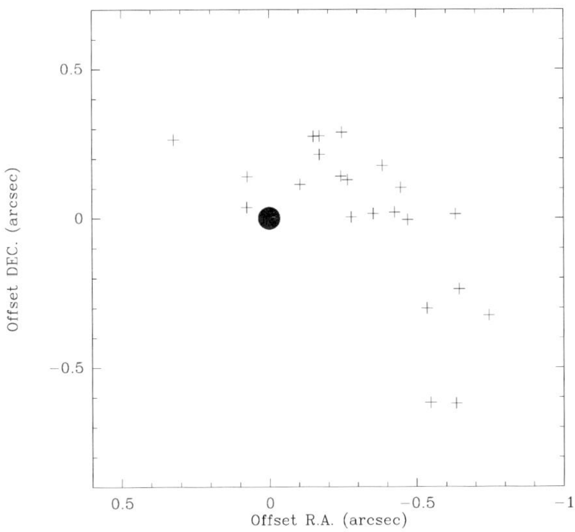

Figure 2. Crosses represent the water maser components. The filled circle represents the position of the 2 μm point source (the reference position of this map).

4. Study of the Surrounding Environment of B1-IRS

4.1. The Central Source

The point-like source detected at 2 μm is located inside the error box of the IRAS source (see Figure 3), at a distance of $\simeq 6''$ from the IRAS catalog position and at $\lesssim 1''$ (350 AU) from the water masers. Due to their proximity, it is very probable that this 2 μm source is the exciting source of the maser emission and of the molecular outflow.

The extended IR emission detected in the K-band is a reflection nebula elongated in the same direction as the water masers.

4.2. The Jet-Like Structure of Water Masers

Our dynamical calculations reveal that motions of the water masers are unbound, since a mass of $\gtrsim 12\ M_\odot$ within 175 AU would be needed to bind them. Such high mass is not consistent with our NH_3 and CCS data. Due to their proximity to the exciting source, and the shape and size of the distribution of the masers, it is very likely that the maser emission delineates the inner ($1''$) part of the jet that is driving the CO outflow, near B1-IRS and detected by Hirano et al. (1997).

4.3. Interaction between CCS and the Molecular Outflow

The velocity pattern observed in the (red-shifted) CCS clumps can be interpreted in terms of infalling clumps, which are strongly interacting with the outflow. What we

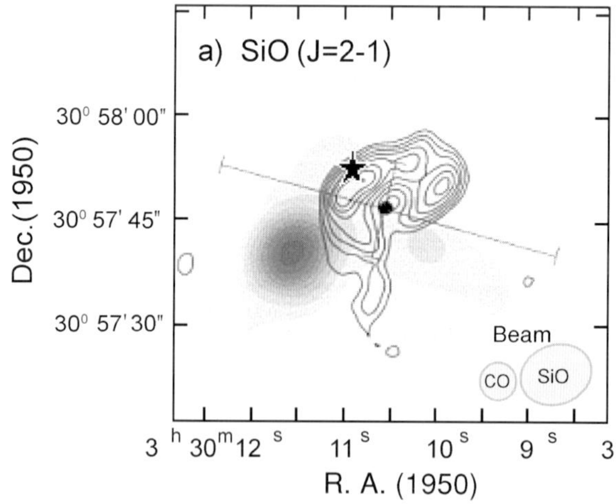

Figure 3. Contours represent the CO(1–0) integrated intensity map from Hirano et al. (1997) superimposed in gray scale on SiO observed by Yamamoto et al. (1992). The black star represents the position of the 2 μm point source. The filled circle marks the position of the IRAS source.

see could be foreground CCS clumps falling toward the central source, but being stopped by the wind, explaining in this way the less red-shifted velocities observed in the CCS lines at positions closer to the central source (Figure 1). However, an alternative possibility is that we could be observing the CCS clumps in the background moving away from the source and being accelerated by the molecular outflow.

5. The Geometry of the CO Molecular Outflow

The moderately high velocity blue-shifted CO emission detected by Hirano et al. (1997) was proposed to be in a near pole-on configuration due to the position of the IRAS source with respect to the outflow, which was located just at the center of the blue-shifted emission (Figure 3).

Our interferometric data and the results of the 2MASS analysis do not support that geometry for the outflow, and we propose a new configuration of the outflow material that fits all our results.

First, the 2 μm point-like source, which is probably powering the outflow (as suggested by its association with the water masers), is not located at the center but at the tip of the blue-shifted CO emission. The extended 2 μm emission is elongated in the same direction that the cluster of water masers, and it is coincident with the shape of the CO outflow. This coincidence further reinforces that the 2 μm source marks the position of the powering source, and that the blue-shifted lobe of the outflow is oriented towards the southwest of it.

Moreover the velocity gradient of the CCS cannot be explained by a pole-on outflow. An outflow in the plane of the sky, however, could sweep the CCS material providing the velocity gradient observed. The geometry on the plane of the sky is reinforced by the lack of a velocity gradient of the water masers along its jet-like structure.

Considering all the available information, an outflow that lies near the plane of the sky with a finite opening angle, can explain all the results obtained with our observations (see Figure 1).

Acknowledgements

GA, IdG, JFG and JMT acknowledge support from MCYT grant (FED-ER funds) AYA2002-00376 (Spain). GA acknowledges support from Junta de Andalucía (Spain). IdG acknowledges the support of a Calvo Rodés Fellowship from the Instituto Nacional de Técnica Aeroespacial and of the summer students program of the National Radio Astronomy Observatory.

Note

1. The NRAO is operated by Associated Universities Inc., under cooperative agreement with the National Science Foundation.

References

Furuya, R.S., Kitamura, Y., Wootten, H.A., Claussen, M.J. and Kawabe, R.: 2001, *ApJ* **559**, L143.
Hirano, N., Kameya, O., Mikami, H., Saito, S., Umemoto, T. and Yamamoto, S.: 1997, *ApJ* **478**, 631.
Kuiper, T.B.H., Langer, W.D. and Velusamy, T.: 1996, *ApJ* **468**, 761.
Rodríguez, L.F., Moran, J.M., Gottlieb, E.W. and Ho, P.T.P.: 1980, *ApJ* **235**, 845.
Saito, S., Kawaguchi, K., Yamamoto, S., Ohishi, M., Suzuki, H. and Kaifu, N.: 1987, *ApJ* **317**, L115.
Suzuki, H., Yamamoto, S., Ohishi, M., Kaifu, N., Ishikawa, S., Hirahara, Y. and Takano, S.: 1992, *ApJ* **392**, 551.
Velusamy, T., Kuiper, T.B.H. and Langer, W.D.: 1995, *ApJ* **451**, L75.
Yamamoto, S., Mikami, H., Saito, S., Kaifu, N., Ohishi, M. and Kawaguchi, K.: 1992, *PASJ* **44**, 459.

THE KINEMATICS, PHYSICAL CONDITION AND MAGNETIC FIELD OF THE W3 IRS5 REGION

HIROSHI IMAI[1] and ANUJ P. SARMA[2]
[1]*Department of Physics, Faculty of Science, Kagoshima University, Japan;*
E-mail: hiroimai@sci.kagoshima-u.ac.jp
[2]*Physics and Astronomy Department, Eastern Kentucky University, U.S.A.*

(Received 16 April 2004; accepted 15 June 2004)

Abstract. We have independently investigated water masers associated with the star-forming region W3 IRS5 with the VLBA. Imai et al., found that the maser 3-D motions exhibit outflows, which likely originate from two of the hyper-compact HII regions in this source. Sarma et al., have detected the Zeeman effect in water masers toward W3 IRS5 and measured line-of-sight magnetic field strengths of between 14 and 42 mG. The directions of maser linear polarization are well aligned in the whole maser region and perpendicular to the estimated magnetic field. These polarimetric results are consistent with an hourglass model of the magnetic field in W3 IRS5. Imai et al., also have analyzed the microstructures found in the individual maser features (~1AU), which exhibit the "fractal fashion" and express turbulence on very small scales. Together, we have demonstrated that observations of water masers enable us to comprehensively enhance our understanding of the early stages of the formation of massive stars in clusters.

Keywords: water masers, kinematics, star formation, outflow, Zeeman effect, linear polarization, clustering, turbulence

1. Introduction

W3 IRS5 is expected to be one of the youngest sites of formation of massive star clusters in the Galaxy because of the existence of water maser emission and hyper compact HII regions. We have comprehensively observed water maser emission in this region and summarized them in this paper (c.f. Imai et al., 2000, 2002, 2003; Sarma et al., 2001). The water maser emission consists of many bright maser features, which are suitable for proper-motion measurements and have high degrees of linear and circular polarization that enable us to measure their Zeeman effect and electric vector position angles (EVPAs). Thus we have obtained a scenario from magnetically super-critical collapse of a molecular cloud core, through formation of massive stars within a short period ($<10^5$, corresponding to duration of water maser excitation), to formation of multiple stellar winds/outflows. The period of such fast evolution of a star-forming region is also indicated by growth degree of turbulence. It is well demonstrated that water maser emission is a powerful tool for elucidating not only kinematics in detail but also for diagnostics of the physical condition and the magnetic field in the very vicinity ($\leq$0.01 pc) of young stellar objects.

2. Observations with the VLBA

The VLBA observations of the W3 IRS5 water masers have been made five times; the first three epochs (March 1997–January 1998) for the maser proper-motion measurement (Imai et al., 2000, 2002), the fourth epoch (November 1998) for linear polarization measurement (Imai et al., 2003), and the fifth epochs (January 2000) for Zeeman effect measurement (Sarma et al., 2001). Except for the fourth epoch, 4-MHz IF channels were divided into 512 spectral channels, yielding a velocity spacing of 0.11 km s^{-1}.

3. Results and Discussion

3.1. The three-dimensional kinematics of W3 IRS5

The proper motions of 108 water maser features exhibited outflows originating from at least two locations, which roughly coincide with those of two of hyper-compact HII regions (Figure 1) (Claussen et al., 1994; Tieftrunk et al., 1997; Imai et al., 2000). This is proved by proper motions of the continuum sources, two of which do not change their relative motions while others exhibit proper motions (Wilson et al., 2003). The directions of the flows are roughly parallel, implying that the massive-star cluster was created during formation of hierarchical structures in a parent molecular cloud core. Through a model fitting analysis assuming an expanding flow with a spherical velocity field and statistical parallax method for the maser kinematics, the distance to W3 IRS5 was estimated to be about 1.8 kpc.

3.2. The magnetic field of W3 IRS5

Zeeman effects found in four of the bright water maser features exhibit a magnetic field strength between 14 and 40 mG; all of the magnetic field vectors point toward an observer (Sarma et al., 2001, 2002). This strength supports a magnetic field–gas density relation observationally founded and theoretically predicted by Crutcher (1999) up to a density of $\sim 10^9$ cm^{-3}, in which water masers are excited.

The EVPAs of water masers do not have correlation with maser proper motions, but alignment in the specific direction ($\sim 75°$ east from north) (Figure 1) (Imai et al., 2003). The EVPAs and the magnetic field strength along the line-of-sight are well explained by an ‘hourglass’ model (e.g., Greaves et al., 1994). The pinch of the hourglass field is located within 5000 AU from the maser region. Thus the

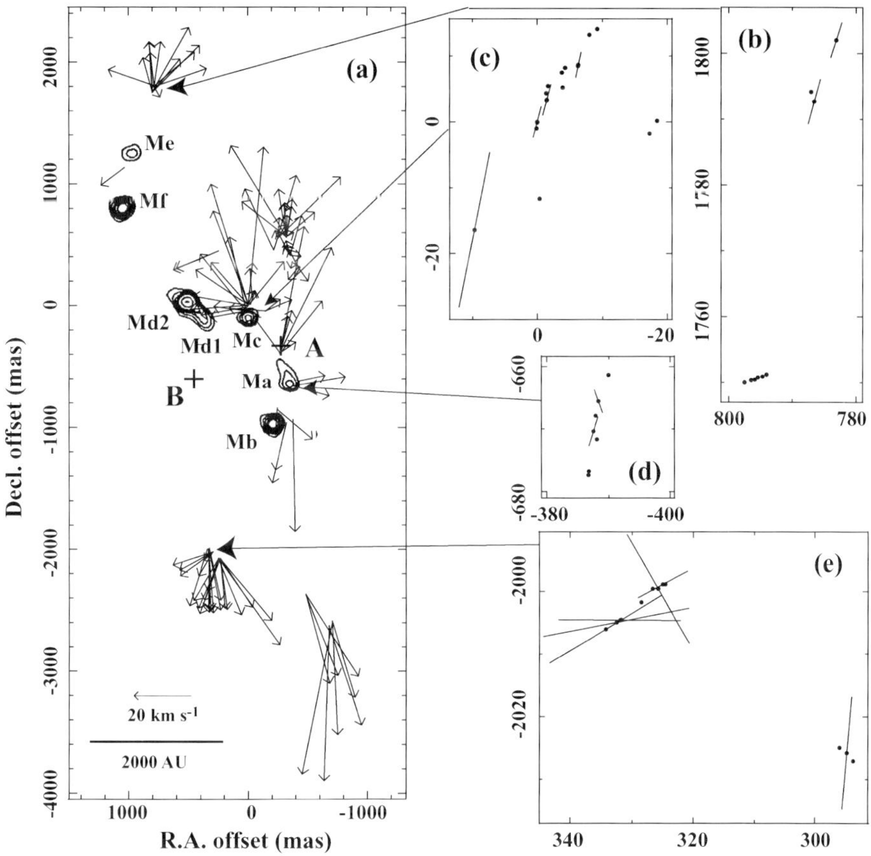

Figure 1. Water masers in W3 IRS5. (a) Proper motions of the water maser features (Imai et al., 2000) and compact continuum sources (Claussen et al., 1994) (contours) in W3 IRS5. Arrows indicate the locations and proper motions of maser features. The designations of the continuum sources are same as those in Tieftrunk et al. (1997). Two plus marks A and B indicate the locations of driving sources of the outflows in W3 IRS5, which are close to the continuum sources Ma and Md1, and were estimated by Imai et al. (2000). (b)–(e) Details of the maser feature distribution and the directions of linear polarization of the maser features (Imai et al., 2003). A length and a position angle of a bar indicate a logarithmic scale of the degree of linear polarization and an EVPA, respectively.

magnetic field in W3 IRS5 may be formed as a result of magnetically supercritical collapse of a molecular cloud core. The collapse enhances the magnetic field at the center of W3 IRS5 by a factor of ~50, which is more enhanced in the maser region by shocks by a factor of ~20. On the other hand, the magnetic field is still smaller than a critical strength ($(8\pi\rho_{\rm maser}v^2_{\rm maser})^{1/2} \simeq 250$ mG); this condition may occur both in the maser region and the ambient region in the outflow. This indicates that the magnetic field cannot control the dynamics of the outflow. The outflow may be driven by strong stellar winds, in which non-thermal continuum emission is ejected such as those found by Wilson et al. (2003).

3.3. Turbulence in W3 IRS5

Individual maser features having a typical size of (∼1 AU) exhibit velocity gradients on a straight line or a complicated shape (see e.g., Imai et al., 2002). Statistical analyses calculating two-point correlation functions exhibit the dimension-less characteristics (fractal fashion) (Figure 2a). Velocity dispersions are also well fit with power–law functions of feature separation in W3 IRS5 in the whole spatial scale (0.01–200 AU) (Figure 2b). It suggests that physical conditions between the interior of maser feature and the larger area are well linked with common turbulence.

Note that the power–law indices of the correlation functions are steeper in W3 IRS5 than those in W49N, the site of more active and later stage of massive star

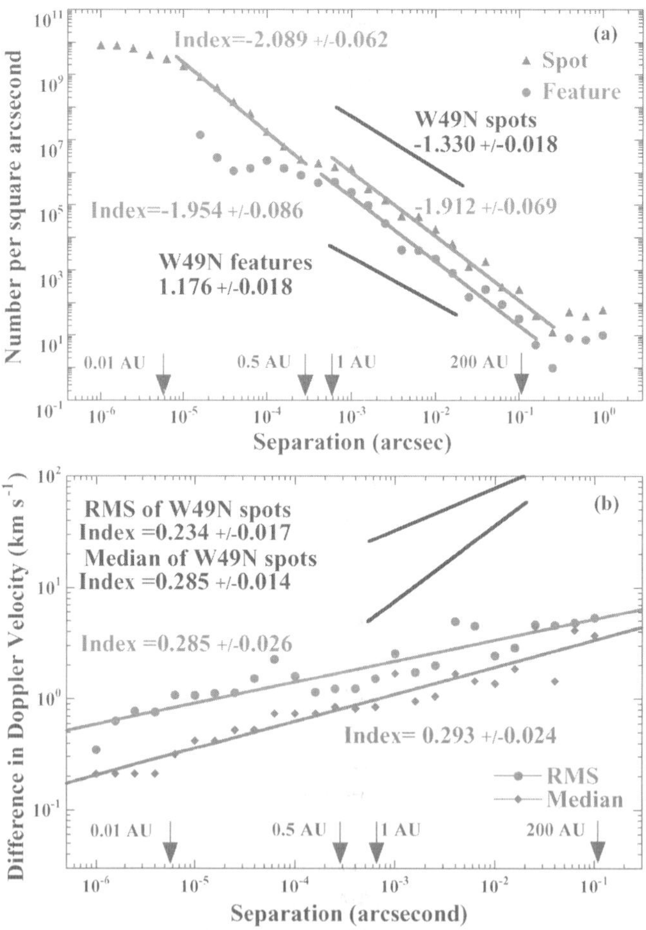

Figure 2. (a) Two-point and (b) velocity correlation functions for the W3 IRS5 water masers (Imai et al., 2002). These functions are compared with those in W49N obtained by Gwinn (1994).

formation (Gwinn et al., 1994). This indicates that the W3 IRS5 water masers have more clumpy structure. Velocity dispersions in W49N suddenly increase (up to 100 km s^{-1}) on a spatial scale larger than 1 AU, indicating turbulence driven by the outflow. It is believed that turbulence is enhanced only at the larger scale and it makes the clumpy structure smoothed so that brighter maser emission is created. Thus, such statistical analyses may be dependent on kinematic energy of outflows or evolutionary status of the star-forming regions (Imai et al., 2002).

Acknowledgements

The NRAO's VLBA is a facility of the National Science Foundation, operated under a cooperative agreement by Associated Universities, Inc.

References

Claussen, M.J., Gaume, R.A., Johnston, K.J. and Wilson, T.L.: 1994, *Astrophys. J.* **424**, L41–L44.
Crutcher, R.M.: 1999, *Astrophys. J.* **520**, 706–713.
Greaves, J.S., Murray, A.G. and Holland, W.S.: 1994, *Astron. Astrophys.* **284**, L19–L22.
Gwinn, C.R.: 1994, *Astrophys. J.* **429**, 241–242.
Imai, H., Horiuchi, S., Deguchi, S. and Kameya, O.: 2003, *Astrophys. J.* **595**, 285–293.
Imai, H., Deguchi, S. Sasao, T.: 2002, *Astrophys. J.* **567**, 971–979.
Imai, H., Kameya, O., Sasao, T., Miyoshi, M., Deguchi, S., Horiuchi, S. and Asaki, Y.: 2000, *Astrophys. J.* **538**, 751–765.
Sarma, A.P., Troland, T.H., Crutcher, R.M. and Roberts, D.A.: 2002, *Astrophys. J.* **580**, 928–937.
Sarma, A.P., Troland, T.H. and Romney, J.D.: 2001, *Astrophys. J.* **554**, L217–L220.
Tieftrunk, A.R., Gaume, R.A., Claussen, M.J., Wilson, T.L. and Johnston, K.J.: 1997, *Astrophys. J.* **318**, 931–946.
Wilson, T.L., Boboltz, D.A., Gaume, R.A. and Megeath, S.T.: 2003, *Astrophys. J.* **597**, 434–442.

EVN OBSERVATIONS OF H_2O MASERS TOWARDS THE HIGH-MASS YOUNG STELLAR OBJECT IN AFGL 5142

C. GODDI[1], L. MOSCADELLI[2], W. ALEF[3] and J. BRAND[4]

[1]*INAF, Osservatorio Astronomico di Cagliari, Loc. Poggio dei Pini, Capoterra (CA), Italy; E-mail: cgoddi@ca.astro.it*

[2]*Dipartimento di Fisica, Università degli Studi di Cagliari, S.P. Monserrato-Sestu Km 0.7, Cagliari, Italy*

[3]*Max-Planck-Institut für Radioastronomie, Auf dem Hügel 69, D-53121 Bonn, Germany*

[4]*Istituto di Radioastronomia CNR, Via Gobetti 101, Bologna, Italy*

(Received 16 April 2004; accepted 15 June 2004)

Abstract. We have conducted multi-epoch EVN observations of the 22.2 GHz water masers towards the high-mass young stellar object in AFGL 5142. With four observing epochs, spanning a time of ~1 year, 12 distinct maser features have been detected and accurate values of the proper motions are derived for those persistent over three or four epochs. On the basis of their spatial distribution, the observed maser features can be divided into two groups. A model fit to the positions and velocities of the maser features of Group I, detected in the same region (within ~500 mas) where the massive YSO should be located, demonstrates that these might arise on the surface of a nearly edge-on Keplerian disk, rotating around a massive young stellar object. The maser features of Group II, found at large distances from the YSO ($\geq 1''$), have positions and line-of-sight velocities in agreement with the blue-shifted lobe of a large-scale molecular outflow, and might result from the interaction between the gas flowing away from the young stellar object and the ambient gas of the progenitor molecular core.

Keywords: masers-stars, formation-ISM, individual objects (AFGL5142)-ISM, kinematics and dynamics-radio lines, ISM

1. Introduction

We present multi-epoch VLBI observations of the 22 GHz H_2O masers towards the massive star forming region AFGL 5142 (IRAS05274 + 3345). Previous observations strongly suggest the presence of a YSO (at RA $= 05^h27^m30^s$; DEC $= +33°45'40''$), coincident with two compact thermal continuum sources (observed with the VLA at 8.4 GHz and with OVRO at 88 GHz), located at the origin of a H_2 and SiO jet and at the center of a CO and HCO^+ outflow (Hunter et al., 1995, 1999).

The flux density of the radio emission (interpreted as free–free emission from an ionized wind) and the bolometric luminosity (estimated using the IRAS fluxes) can be explained if the exciting source is a massive object, with spectral type B2 or earlier (Hunter et al., 1999). VLA NH_3-observations (Zhang et al., 2002) show a compact structure (diameter ~1800 AU), coincident in position with the 8.4 and 88 GHz continuum sources, interpreted as an unresolved rotating disk.

The 22.2 GHz water masers in AFGL 5142 have been observed with the VLA at three epochs (1991, Torrelles et al., 1992; 1992, Hunter et al., 1995, 1998, 1999), and found to be distributed within a few arcseconds from the 8.4 and 88 GHz continuum sources. However, the VLA angular resolution ($\sim 0.''1$) is inadequate to determine the detailed spatial distribution of the maser spots, to measure their proper motions and, consequently, to investigate the kinematics traced by the H_2O masers in this source.

2. Observations and Results

AFGL 5142 was observed in the (6_{16}-5_{23}) H_2O maser line (rest frequency 22235.080 MHz) using the European VLBI Network (EVN) at four epochs (October 1996, and June, September, November 1997). The data were processed with the MKIII correlator at the Max–Planck-Institut für Radioastronomie (Bonn, Germany), obtaining 112 spectral channels with a separation of 0.12 km s^{-1}. Data reduction was performed using the NRAO AIPS package, following the standard procedure for VLBI line data.

We have produced channel maps extended over a sky area of ($\Delta\alpha \cos\delta \times \Delta\delta$)$4'' \times 4''$ and covering a radial velocity range of 8 km s^{-1}. The CLEAN beam was an elliptical gaussian with a typical FWHM size of $\sim$1 mas. Each channel map was searched for emission above a conservative detection threshold (in the range 5–10 σ, where $\sigma = 0.03$–0.3 Jy/beam), and the detected maser spots were fitted with two-dimensional elliptical Gaussians, determining position, flux density, and FWHM size of the emission. Hereafter, we use the term of maser "feature" to indicate a group of maser spots detected in at least three contiguous channels, with a position shift of the intensity peak from channel to channel smaller than the FWHM size.

Counting all four epochs, 26 H_2O maser features were detected. Several of these show a good agreement in relative positions (within few mas) and line-of-sight velocities (within 1 km s^{-1}) for two or more epochs, and therefore we assume that in these cases we identified features which persist over time.

Figure 1 compares our VLBI results with previous interferometric observations. Top panels show the HCO^+ and SiO outflows observed with OVRO (Hunter et al., 1999). The area where the 22.2 GHz water maser emission has been detected with our VLBI observations is individuated by a small rectangle at the center of the field of view. This area is expanded in the lower panel of Figure 1, which shows the spatial distribution of the VLBI maser features superimposed on top of the distribution of the VLA emission centers found by Hunter et al. (1995, 1999) at two different epochs.

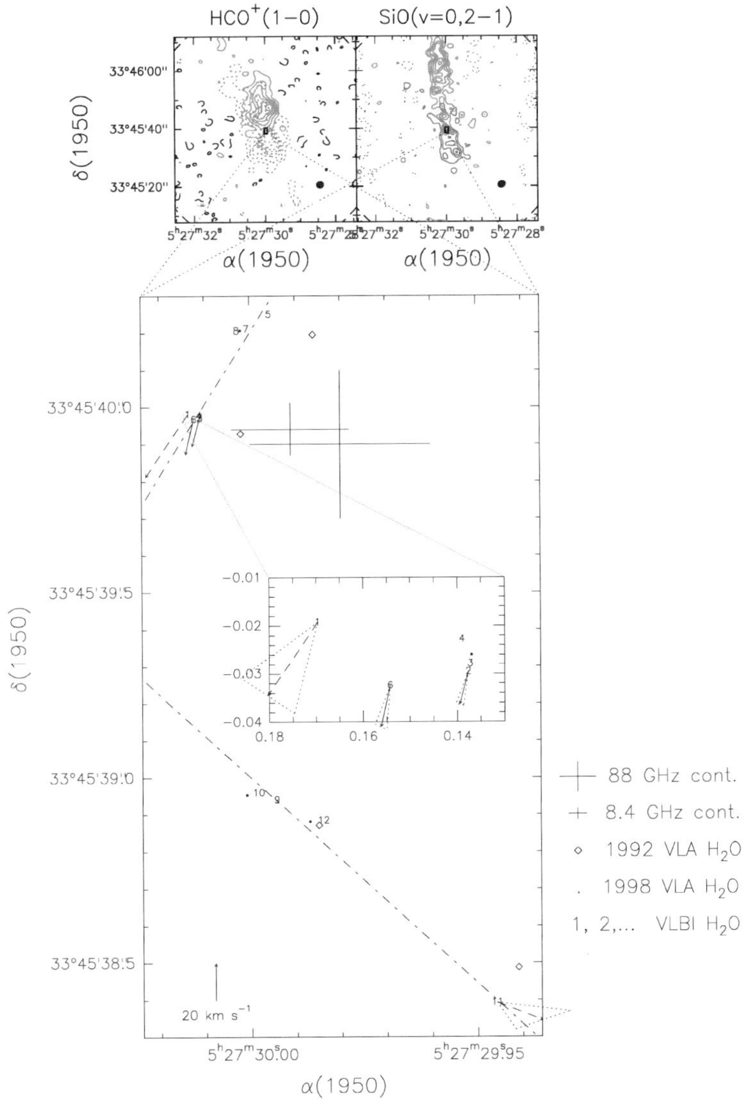

Figure 1. The upper panels show the contour maps of the OVRO high-velocity molecular line emission for HCO^+ ($1 \rightarrow 0$) (*left*) and SiO ($v = 0, 2 \rightarrow 1$) (*right*). The area at the center of the OVRO field of view, where the 22.2 GHz water emission has been detected by our VLBI observations, is expanded in the lower panel. This shows the spatial distribution of the VLBI features superimposed on top of the distribution of the VLA emission centers. The arrows indicate the measured proper motions, with dashed lines used in case of more uncertain values. The dotted triangles drawn around the proper motion vectors represent the amplitude and orientation uncertainty. The region in the northeast corner enclosed by the dotted lines is shown in greater detail in the box at the center of the lower panel, whose coordinates are arcsec offsets relative to the position: RA = $05^h27^m30^s$; DEC = $+33^\circ45'40''$. The amplitude scale for the proper motions is given at the bottom left corner of the panel. For each of the two Groups of VLBI maser features, the dot-dashed line indicates the axis whose average distance from the feature positions is minimum.

The proper motions (indicated by the arrows in the Figure 1) have been calculated performing a linear least-squares fit of the positional offsets with time. Among the features observed at three or more epochs, the proper motions are derived only for those moving in a straight line at constant velocity. Tentative values of the proper motions (indicated by dashed arrows) are calculated also for the features observed at only two epochs.

The absolute position of the VLBI map has been determined by shifting it on top of the VLA map and minimizing the root mean square difference between the positions of the VLBI and VLA maser emission centers. Founding a good overlap between them, we are confident that the absolute position derived for the VLBI map is accurate within the VLA positional uncertainty.

On the basis of their spatial distribution, the VLBI maser features can be divided into two groups. Group I is found in the northeast corner of the area plotted in the lower panel of Figure 1, in the same region (within ~500 mas) where the 8.4 GHz and 88 GHz continuum emissions are detected. Group II is more detached ($\geq 1''$) southward from the continuum emissions.

3. Discussion

The two groups of maser features have well distinct spatial and velocity distributions.

The maser features of Group I have a sky-projected distance $\leq$500–1000 AU from the 8.4 GHz and 88 GHz continuum sources, and should emerge near to the expected location of the massive YSO. At such a close distance to the YSO, the current theory of star formation predicts that an accretion disk and/or the base of the associated jet should be found. Both these environments might offer suitable physical conditions (high densities, $n_{\mathrm{H_2}} \sim 10^7\ \mathrm{cm}^{-3}$, and relatively high temperatures, $T \geq 400$ K) for the excitation of the 22.2 GHz water masers. Looking at Figure 1, one notes that the maser features of Group I have an elongated spatial distribution (the dot-dashed line indicates the elongation axis) and that the measured proper motions have an orientation close to that of the elongation axis. This geometrical condition is what one would in principle expect if these maser features traced a rotating disk seen edge-on or a collimated outflow motion along the disk axis. To establish more quantitatively which of these kinematical structures is compatible with our measurements, we fitted the position and the velocities of the maser features of Group I with two simple models: (1) a Keplerian disk; (2) a conical outflow.

As only the Keplerian disk model produces a solution for which the measured transversal velocities of the maser features are consistent with the corresponding line-of-sight velocities, we prefer it to the alternative conical flow model. The best fit solution is found with the disk seen almost edge-on and oriented on the sky parallel to the elongation axis of the Group I features (at P.A. $= 153^\circ$). The fitted value of the

YSO mass, $M_{YSO} = 38\,M_{\odot} \pm 20\,M_{\odot}$, although determined with high uncertainty, strongly indicates that the central object is a massive YSO ($M > 10\,M_{\odot}$), consistent with the picture of an exciting object of spectral type B2 or earlier (Hunter et al., 1999). This result is also in agreement with previous mass estimates: the mass of the dusty core associated with the 88 GHz emission (Hunter et al., 1999) and the "disk" mass estimated from the NH_3 measurements (Zhang et al., 2002). Adopting a distance of 1.8 kpc to AFGL 5142, the range of disk radii traced in our model by the maser emission extends from ~30 AU to ~800 AU, which is consistent with the size of several hundreds of AU expected for an accreting disk around a massive YSO.

Owing to the large distance (2000–3000 AU) from the YSO, the Group II of maser features might also be associated with a distinct (as yet undetected) YSO. However, we found that the spatial and velocity distribution of these features is qualitatively in agreement with the blue-shifted lobe of the molecular outflow traced, on much larger scales, by the HCO^+ and SiO emission, suggesting that they might be tracing the flow motion in the innermost portion of the molecular outflow.

In conclusion, our 22 GHz water maser VLBI observations reinforces the idea that AFGL 5142 is a good candidate of high-mass YSO. The interpretation of the VLBI data is based on a scenario where the maser emission emerges both from a keplerian disk and from a jet/outflow system.

References

Hunter, T., Testi, L., Taylor, G., Tofani, G., Felli, M. and Phillips, T.: 1995, *A&A* **302**, 249.

Hunter, T.R., Testi, L., Zhang, Q. and Sridharan, T.K.: 1999, *AJ* **118**, 477–487.

Torrelles, J.M., Gomez, J.F., Anglada, G., Estalella, R., Mauersberger, R. and Eiroa, C.: 1992, *ApJ* **392**, 616–621.

Zhang, Q., Hunter, T., Sridharan, T. and Ho, P.: 2002, *ApJ* **566**, 982.

AGN/STARBURST EVOLUTION

GRAVITATIONALLY INDUCED INFLOW IN STARBURSTS AND AGN

J.H. KNAPEN

Centre for Astrophysics Research, University of Hertfordshire, Hatfield, Herts AL10 9AB, UK; E-mail: j.hknapen@star.herts.ac.uk

(Received 16 April 2004; accepted 15 June 2004)

Abstract. There is considerable evidence that the circumnuclear regions of galaxies are intimately related to their host galaxies, most directly through their bars. There is also convincing evidence for relations between the properties of supermassive black holes in the nuclei of galaxies and those of their host galaxies. It is much less clear, however, how stellar (starburst) and non-stellar (AGN) activity in the nuclear regions can be initiated and fuelled. Here, we review the evidence for gas transport from the disk to the nuclear and circumnuclear regions of galaxies, as well as the statistical relationships between the occurrence of nuclear activity and mechanisms which can cause central gas concentration. In particular, I explore the roles played by bars and interactions, and conclude that in specific, mostly extreme, cases bars and interactions are indeed observed to be connected to nuclear activity. The overall lack of observational evidence for direct causal relationships between the presence of bars and interactions on the one hand, and starburst or Seyfert activity on the other could, however, easily be due to the possibility that we are not yet considering the correct spatial- or time-scales.

Keywords: galaxies: kinematics and dynamics, galaxies: spiral, galaxies: structure, galaxies: active, galaxies: starburst

1. Introduction

Black holes are ubiquitous in the nuclei of both active and non-active galaxies (e.g., Kormendy and Richstone, 1995), and are thought to be at the direct origin of non-stellar nuclear activity (e.g., Lynden-Bell, 1969; Begelman et al., 1984). The fact that the mass of the central supermassive black hole (SMBH) in a galaxy is correlated with the velocity dispersion of the bulge, and hence with its mass (Ferrarese and Merritt, 2000; Gebhardt et al., 2000), provides the most tangible link between the nuclear regions and their host galaxies. But because not all galaxies with SMBHs have AGN characteristics, the presence of an SMBH in itself cannot be enough to make a galaxy "active", at least not continuously.

In the case of starburst galaxies, defined rather loosely as galaxies which show abnormally enhanced massive star formation activity in their central regions (or in some more extreme cases throughout the galaxy), a similar question can be posed, namely what turns the starburst on. In both the AGN and the starbursts, it is the availability of fuel at the right place and the right time which must play a critical role. Such gaseous fuel is plentiful in the disks of galaxies, but must lose significant amounts of angular momentum in order to move radially inward. Large stellar bars, as well as tidal interactions and mergers, have been identified as prime candidates to drive gas efficiently from the outer disk into the inner kiloparsec (see next section). In this short review we will concentrate on the observational evidence, mostly statistical in nature, for the effectiveness of these gravitational mechanisms, concentrating on the effects of bars in Section 2, and on those of interactions in Section 3. Related reviews considering the fuelling of primarily AGN include those by Shlosman et al. (1990), Beckman (2001), Shlosman (2003), Wada (2004), and Jogee (2004).

2. The Effects of Bars

Theoretically and numerically, bars are expected to concentrate gas in the central regions of spiral galaxies because the torqued and shocked gas within the bar loses angular momentum which allows the gas to move further in (e.g., Schwarz, 1984; Combes and Gerin, 1985; Noguchi, 1988; Shlosman et al., 1989; Knapen et al., 1995a). Pieces of observational evidence to fit this picture have been published over the years, both from observations of gas tracers in barred and non-barred galaxies (e.g., Sakamoto et al., 1999; Jogee et al., 2004; Sheth et al., 2004), and from other, less direct, measures of the gas concentration (e.g., Maiolino et al., 1999; Alonso-Herrero and Knapen, 2001). These results have been reviewed in somewhat more detail by Knapen (2004), and we limit ourselves here to the conclusion that there is increasing observational support for the theoretical suggestion that bars lead to gas accumulation in the central regions of galaxies. Whether this accumulation is related to the occurrence of nuclear activity of the non-stellar or stellar variety is not known, though, and a careful consideration of both spatial- and timescales must be made to connect gravitationally driven inflow with fuelling of the starburst and/or the AGN. In the remainder of this section, we will review the question of whether there is observational evidence that bars are related to the occurrence of nuclear activity.

2.1. Bars and Starburst Activity

There is a clearly observed trend for nuclear starbursts to occur preferentially in barred hosts (e.g., Hummel, 1981; Hawarden et al., 1986; Devereux, 1987; Dressel, 1988; Puxley et al., 1988; Arsenault, 1989; Huang et al., 1996; Martinet and Friedli, 1997; Hunt and Malkan, 1999; Roussel et al., 2001). For example, Hummel (1981) reported that the central radio continuum component is typically twice as strong in barred as in non-barred galaxies; Hawarden et al. (1986) found that barred galaxies dominate the group of galaxies with a high 25 μm/12 μm flux ratio; and Arsenault (1989) found an enhanced bar plus ring fraction among starburst hosts. Huang et al. (1996) revisited IRAS data to confirm that starburst hosts are preferentially barred, but did point out that this result only holds for strong bars (SB class in the RC3 catalogue, de Vaucouleurs et al., 1991) and in early-type galaxies, results confirmed more recently by Roussel et al. (2001). In contrast, Isobe and Feigelson (1992) did not find an enhanced far-IR to blue flux ratio among barred galaxies, and Ho et al. (1997) found only a very marginal increase in the detection rate of H II nuclei (indicative of starburst activity) among the barred as compared to non-barred galaxies in their sample, only among the late-type spirals (Sc-Sm), and most likely as the result of selection effects rather than bar-induced inflow (Ho et al., 1997). All results mentioned above rely on optical catalogues such as the RC3 to derive the morphological classifications, whereas it is now well-known that the presence of a bar can be deduced more reliably from near-IR imaging (e.g., Scoville et al., 1988; Knapen et al., 1995b). Although near-IR imaging leads to enhanced bar fractions as compared to optical imaging (e.g., Knapen et al., 2000; Eskridge et al., 2000), it is not clear how it would affect the results on bars and starbursts.

The statistical studies referred to above thus seem to show that bars and starbursts are connected, but that the results are subject to important caveats and exclusions. Further study is needed, determining bar parameters from near-IR imaging, using carefully defined samples, and exploring more direct starburst indicators than the IRAS fluxes which have often been used. Higher-resolution imaging of the starburst galaxies is also needed, to confirm the possible *circum*nuclear nature of the starburst, already suggested back in 1986 by Hawarden et al.

2.2. Bars and Seyfert Activity

Seyferts are almost ideal for a study of AGN host galaxies: they are relatively local and occur predominantly in disk galaxies (see Figure 1 for a nice example). One of the aspects of the host galaxy—Seyfert activity connection that has received a good deal of attention over the years is the question of whether Seyfert hosts are more often barred than non-Seyferts. Starting with the work of Adams (1977), many authors have dedicated efforts to resolve this question, without finding conclusive evidence (e.g., Adams, 1977; Simkin et al., 1980; Balick and Heckman, 1982; MacKenty, 1990; Moles et al., 1995; Ho et al., 1997; Crenshaw et al., 2003). Unfortunately, many of these investigations are plagued by the absence of a properly matched control sample, by the use of the RC3 classification or, worse, ad-hoc and non-reproducible classification criteria to determine whether a galaxy is barred; and all of them are based on optical imaging.

Near-IR imaging is much better suited for finding bars (see Section 2.1), and a small number of studies have combined the use of high-quality, near-IR imaging with the careful selection and matching of samples of Seyfert and quiescent galaxies. One such study, by Mulchaey and Regan (1997), reports identical bar fractions, but Knapen et al. (2000), using imaging at higher spatial resolution and a rigorously applied set of bar criteria, find a marginally significant difference, with a higher bar fraction in a sample of CfA Seyferts than in a control sample of non-Seyferts

More in general, and considering galaxies less extreme than those in the ULIRG class, there is considerable evidence for a connection between interactions and enhanced star formation in galaxies, often measured using galaxy colours which are bluer in the case of current star formation (see, e.g., the seminal paper by Larson and Tinsley, 1978). But even so, a more detailed consideration can expose important caveats. We mention the recent paper by Bergvall et al. (2003), who considered two matched samples of nearby interacting (pairs and clear cases of mergers) and non-interacting galaxies, and measured star formation indices based on UBV colours. From this analysis, Bergvall et al. do *not* find evidence for significantly enhanced star-forming activity among the interacting/merging galaxies, although they do report a moderate increase in star formation in the very centres of their interacting galaxies. Interesting in this respect are also recent results from a combination of Sloan Digital Sky Survey and 2dF Galaxy Redshift Survey data, presented by Balogh et al. (2004). These authors study the equivalent width of Hα emission and find no correlation between its distribution among the star-forming population and the environment.

So although mergers can undoubtedly lead to massive starbursts, they appear to do so only in rare cases. Bergvall et al. (2003) estimate that only about 0.1% of a magnitude limited sample of galaxies will host massive starbursts generated by interactions and mergers. Most interactions between galaxies may not lead to a notable increase in the starburst activity. Those that do may be selected cases where a set of parameters, both internal to the galaxies and regarding the orbital geometry of the merger, is conducive to the occurrence of starburst activity (e.g., Mihos and Hernquist, 1996). To further illustrate this point, we quote the results published by Laine et al. (2003), who find very little evidence for trends in starburst activity from detailed *HST* imaging of the Toomre sequence of merging galaxies.

3.2. Interactions and Seyfert Activity

Seyfert activity is well known to occur in interacting and merging galaxies, and several rather spectacular examples are well known (for instance NGC 2992, or a number of the closest ULIRGs such as Mrk 273). To check statistically whether there is a connection between interactions and the occurrence of nuclear activity, authors have considered the numbers of companions to Seyfert galaxies as compared to non-active control galaxies (e.g., Fuentes-Williams and Stocke, 1988; de Robertis et al., 1998; Schmitt, 2001), or, alternatively, have searched for different fractions of Seyfert or AGN activity among more or less crowded environments (e.g., Kelm et al., 1998; Miller et al., 2003). The conclusion from this substantial body of work must be that no unambiguous evidence exists for a direct connection between the occurrence of Seyfert activity and interactions. Some earlier work did report claims of a statistical connection, but was unfortunately plagued by poor control sample selection (see Laurikainen and Salo, 1995 for a detailed review), and most early studies, but also some of the recent ones, are not based on complete sets of redshift information for the possible companion galaxies. In addition, Laine et al. (2002) have shown that the bar fraction among both the Seyfert and non-Seyfert galaxies in their sample is completely independent of the presence of faint, or bright, companions (interacting galaxies were not considered by Laine et al.). We thus conclude that interactions and Seyfert activity may well be linked in individual cases, but that as yet the case that they are statistically connected has not been made convincingly.

4. Concluding Remarks

Although it is evident from theory and modelling, and increasingly also from observations, that bars can drive gaseous material to the central kiloparsec-scale regions of a galaxy (Section 2), the evidence that this centrally condensed gas leads to AGN or starburst activity remains rather elusive. Starbursts are provoked by bars or interactions, at least in some cases, though most probably not in general (Sections 2.1 and 3.1). There is some evidence that Seyfert activity preferentially occurs in barred host galaxies, but the effect is only a statistical one, and not very strong either (Section 2.2). There seems to be no evidence that Seyfert hosts are more often interacting than non-Seyferts (Section 3.2).

Since we have known for quite some time that net radial gas inflow must be accompanied by the loss of significant amounts of angular momentum, and that the kind of deviations from axisymmetry in the gravitational potential of the host galaxy set up by bars and interactions is well suited to lead to such angular momentum loss (see reviews by Shlosman et al., 1989, 1990; Shlosman, 2003), we must be missing some part of the puzzle. One possibility is that we are not looking at the right things at the right time: the spatial- as well as the time-scales under consideration may not be correct. So far, the spatial scale considered observationally has been from tens of kiloparsec down to, say, a hundred parsec. Whereas this range may be wholly adequate for the study of a major starburst, which can span up to a kiloparsec, it may well be wholly *in* adequate for AGN fuelling, which is expected to be related to accretion to a SMBH, on scales of AUs. As far as timescales are concerned, what has been considered in the studies reviewed here is generally a rather long-lived phenomenon influencing kiloparsec-scale regions (bar or galaxy-galaxy interaction). Starburst or AGN activity, on the other hand, occurs on essentially unknown (e.g., for Seyferts) or rather short (around 10^7 years could be expected for most starbursts) timescales. If the starburst or AGN activity is indeed short-lived, and possibly also periodic, the connection between the presence of activity at the currently observed epoch and any parameter of the host galaxy is not necessarily straightforward (as pointed out, e.g., by Beckman, 2001).

The fact that we see any correlations at all, such as those of bar fractions with the presence of starburst or Seyfert activity, then, tells us that bars and interactions do have some role to play, as we would have expected in any case from theory and modelling. In the coming years, we must start to disentangle the effects of gravitationally induced gas inflow, which brings gas to the inner kiloparsec region at least, from other possible mechanisms which can transport the gas further in, and from the time scales and duty cycles of the activity. Given that purely morphological studies of the central regions of active and non-active galaxies seem to have reached their limits (e.g., Laine et al., 2002), one must move on to study the gas and stellar kinematics and dynamics.

Integral field spectroscopy (e.g., Bacon et al., 2001), especially when used in conjunction with adaptive optics techniques, should allow a good deal of progress here, giving simultaneous high-resolution mapping of the distributions of stellar populations and dust, as well as of the gas and stellar kinematics. In combination with detailed numerical modelling, this could lead to the detection of the dynamical effects of, e.g, nuclear bars on gas flows which may be more directly related to the fuelling process of starbursts and/or AGN.

Acknowledgements

I am indebted to my collaborators on the various aspects of the work described here, especially Shardha Jogee, Seppo Laine, Reynier Peletier, and Isaac Shlosman. The William Herschel Telescope is operated on the island of La Palma by the Isaac Newton Group in the Spanish Observatorio del Roque de los Muchachos of the Instituto de Astrofí sica de Canarias.

References

Adams, T.F.: 1977, *ApJS* **33**, 19.
Alonso-Herrero, A. and Knapen, J.H.: 2001, *AJ* **122**, 1350.
Armus, L., Heckman, T. and Miley, G.: 1987, *AJ* **94**, 831.
Arsenault, R.: 1989, *A&A* **217**, 66.
Bacon, R., et al.: 2001, *MNRAS* **326**, 23.
Balick, B. and Heckman, T.M.: 1982, *ARA&A* **20**, 431.
Balogh, M. et al.: 2004, *MNRAS* **348**, 1355.
Beckman, J.E.: 2001, in J.H. Knapen, J.E. Beckman, I. Shlosman and T.J. Mahoney (eds.), *ASP Conf. Ser. 249, The Central Kiloparsec of Starbursts and AGN: The La Palma Connection*, Astronomical Society of the Pacific, San Francisco, p. 11.
Begelman, M.C., Blandford, R.D. and Rees, M.J.: 1984, *Rev. Mod. Phys.* **56**, 255.
Bergvall, N., Laurikainen, E. and Aalto, S.: 2003, *A&A* **405**, 31.
Clements, D.L., Sutherland, W.J., McMahon, R.G. and Saunders, W.: 1996, *MNRAS* **279**, 477.
Combes, F. and Gerin, M.: 1985, *A&A* **150**, 327.
Crenshaw, D.M., Kraemer, S.B. and Gabel, J.R.: 2003, *AJ* **126**, 1690.
de Robertis, M.M., Yee, H.K.C. and Hayhoe, K.: 1998, *ApJ* **496**, 93.
de Vaucouleurs, G., de Vaucouleurs, A., Corwin, J.R., Buta, R.J., Paturel, G. and Fouque, P.: 1991, Third reference catalogue of Bright galaxies, 1991, Springer-Verlag (RC3), New York.
Devereux, N.: 1987, *ApJ* **323**, 91.
Dressel, L.L.: 1988, *ApJ* **329**, L69.

Erwin, P. and Sparke, L.S.: 2002, *AJ* **124**, 65.
Eskridge, P.B. et al.: 2000, *AJ* **119**, 536.
Eskridge, P.B. et al.: 2002, *ApJS* **143**, 73.
Ferrarese, L. and Merritt, D.: 2000, *ApJ* **539**, L9.
Fuentes-Williams, T. and Stocke, J.T.: 1988, *AJ* **96**, 1235.
Gebhardt, K. et al.: 2000, *ApJ* **539**, L13.
Genzel, R. et al.: 1998, *ApJ* **498**, 579.
Hawarden, T.G., Mountain, C.M., Leggett, S.K. and Puxley, P.J.: 1986, *MNRAS* **221**, 41P.
Ho, L.C., Filippenko, A.V. and Sargent, W.L.W.: 1997, *ApJ* **487**, 591.
Huang, J.H., Gu, Q.S., Su, H.J., Hawarden, T.G., Liao, X.H. and Wu, G.X.: 1996, *A&A* **313**, 13.
Hummel, E.: 1981, *A&A* **93**, 93.
Hunt, L.K. and Malkan, M.A.: 1999, *ApJ* **516**, 660.
Isobe, T. and Feigelson, E.D.: 1992, *ApJS* **79**, 197.
Jogee, S. 2004, in D. Alloin, R. Johnson and P. Lira (eds.), *AGN Physics on All Scales*, Springer, Berlin, in press.
Jogee, S., Scoville, N.Z. and Kenney, J.: 2004, *ApJ* in press (astro-ph/0402341).
Joseph, R.D. and Wright, G.S.: 1985, *MNRAS* **214**, 87.
Kelm, B., Focardi, P. and Palumbo, G.G.C.: 1998, *A&A* **335**, 912.
Knapen, J.H.: 2004, in *The Neutral ISM in Starburst Galaxies*,. S. Aalto, S. Hüttemeister, and A. Pedlar (eds.), Astronomical Society of the Pacific: San Francisco, in press (astro-ph/0312172).
Knapen, J.H., Beckman, J.E., Heller, C.H., Shlosman, I. and de Jong, R.S.: 1995a, *ApJ* **454**, 623.
Knapen, J.H., Beckman, J.E., Shlosman, I., Peletier, R.F., Heller, C.H. and de Jong, R.S.: 1995b, *ApJ* **443**, L73.
Knapen, J.H., Shlosman, I. and Peletier, R.F.: 2000, *ApJ* **529**, 93.
Knapen, J.H., de Jong, R.S., Stedman, S. and Bramich, D.M.: 2003, *MNRAS* **344**, 527 (Erratum MNRAS 346, 333).
Kormendy, J. and Richstone, D.: 1995, *ARA&A* **33**, 581.
Laine, S., Shlosman, I., Knapen, J.H. and Peletier, R.F.: 2002, *ApJ* **567**, 97.
Laine, S., van der Marel, R.P., Rossa, J., Hibbard, J.E., Mihos, J.C., Böker, T. and Zabludoff, A.I.: 2003, *AJ* **126**, 2717.
Larson, R.B. and Tinsley, B.M.: 1978, *ApJ* **219**, 46.
Laurikainen, E. and Salo, H.: 1995, *A&A* **293**, 683.
Laurikainen, E., Salo, H. and Rautiainen, P.: 2002, *MNRAS* **331**, 880.
Laurikainen, E., Salo, H., and Buta, R. 2004, *ApJ* **607**, 103.
Lynden-Bell, D.: 1969, *Nature* **223**, 690.
MacKenty, J.W.: 1990, *ApJS* **72**, 231.
Maiolino, R., Risaliti, G. and Salvati, M.: 1999, *A&A* **341**, L35.
Martinet, L. and Friedli, D.: 1997, *A&A* **323**, 363.
Mihos, J.C. and Hernquist, L.: 1996, *ApJ* **464**, 641.
Miller, C.J., Nichol, R.C., Gómez, P.L., Hopkins, A.M. and Bernardi, M.: 2003, *ApJ* **597**, 142.
Moles, M., Márquez, I. and Pérez, E.: 1995, *ApJ* **438**, 604.
Mulchaey, J.S. and Regan, M.W.: 1997, *ApJ* **482**, L135.
Murphy, T.W., Armus, L., Matthews, K., Soifer, B.T., Mazzarella, J.M., Shupe, D.L., Strauss, M.A. and Neugebauer, G.: 1996, *AJ* **111**, 1025.
Noguchi, M.: 1988, *A&A* **203**, 259.
Puxley, P.J., Hawarden, T.G. and Mountain, C.M.: 1988, *MNRAS* **231**, 465.
Roussel, H. et al.: 2001, *A&A* **372**, 406.
Sakamoto, K., Okumura, S.K., Ishizuki, S. and Scoville, N.Z.: 1999, *ApJ* **525**, 691.
Sanders, D.B., Soifer, B.T., Elias, J.H., Madore, B.F., Matthews, K., Neugebauer, G. and Scoville, N.Z.: 1988, *ApJ* **325**, 74.

Sanders, D.B. and Mirabel, I.F.: 1996, *ARA&A* **34**, 749.
Schmitt, H.R.: 2001, *AJ* **122**, 2243.
Schwarz, M.P.: 1984, *MNRAS* **209**, 93.
Scoville, N.Z., Matthews, K., Carico, D.P. and Sanders, D.B.: 1988, *ApJ* **327**, L61.
Sheth, K., Vogel, S.N., Regan, M.W., Teuben, P.J., Harris, A.I., Thornley, M.D. and Helfer, T.T.: 2004, *ApJ* submitted.
Shlosman, I.: 2003, in S. Collin, F. Combes and I. Shlosman (eds.), *ASP Conf. Ser. 290, Active Galactic Nuclei: From Central Engine to Host Galaxy*, Astronomical Society of the Pacific, San Francisco, p. 427.
Shlosman, I., Frank, J. and Begelman, M.C.: 1989, *Nature* **338**, 45.
Shlosman, I., Begelman, M.C. and Frank, J.: 1990, *Nature* **345**, 679.
Shlosman, I., Peletier, R.F. and Knapen, J.H.: 2000, *ApJ* **535**, L83.
Simkin, S.M., Su, H.J. and Schwarz, M.P.: 1980, *ApJ* **237**, 404.
Wada, K.: 2004, in L.C. Ho (ed.), *Coevolution of Black Holes and Galaxies*, Cambridge University Press, Cambridge, p.187.

FEEDING MONSTERS – A STUDY OF ACTIVE GALAXIES

MELANIE KRIPS[1], ANDREAS ECKART[1], ROBERTO NERI[2], JOERG-UWE POTT[1,3], JENS ZUTHER[1], JULIA SCHARWÄCHTER[1] and THOMAS BERTRAM[1]

[1] *I. Physikalisches Institut, University of Cologne, Germany; E-mail: Krips@ph1.uni-koeln.de*
[2] *IRAM, Grenoble, France*
[3] *ESO, Garching, Germany*

(Received 16 April 2004; accepted 15 June 2004)

Abstract. We present new interferometric observations of the molecular gas distribution and kinematics in the nuclei of different active galaxies at high angular resolution and high sensitivity carried out with the IRAM Plateau de Bure interferometer (PdBI). The observations cover galaxies in a redshift range of ~0.03–1.4. We have so far observed five different active galaxies: NGC3718, NGC1068, HE1029–1831, 3C48 and Q0957 + 561. The first two objects belong to the NUGA (NUclei of GAlaxies) project – an international collaboration mainly between Spain, France and Germany – containing about 30 Seyfert and LINER galaxies. Both are at the same distance of ~14Mpc and show a warped gas and dust disk. The new mosaic map of NGC3718 corrected for short-spacing effects with IRAM 30 m observations well demonstrates the existence of the warped gas disk with several secondary maxima in the projected gas distribution caused by orbit crowding effects. Based on these new data we have improved recent simulations of the warped disk in NGC3718. HE1029–1831 and 3C48 are nearby QSOs. HE1029–1831 is taken out of the Cologne nearby QSO sample. The maps of the integrated CO(1–0) and CO(2–1) emission clearly show that the molecular gas is mostly related to the central bulge with a non-negligible fraction distributed along the bar known from optical observations. Our new CO data of 3C48 unveil new information about the kinematics of its molecular gas complementing and improving further studies based on previous infrared observations and detailed multi-particle simulations. Finally, new measurements of the CO(1–0) line in Q0957 + 561 – a highly-red-shifted, gravitationally lensed quasar – will be presented as a link to earlier evolutionary stages of active galaxies and their hosts.

Keywords: active galaxies, galaxies: kinematics and dynamics, quasars, LINERs

1. Introduction

The phenomenon of active galactic nuclei remains an intensively debated subject. Knowledge of the distribution and kinematics of circumnuclear molecular gas in this type of objects is a key issue for the understanding of the mechanisms fueling the central engine with enough matter supposedly responsible for the observed high activity. In contrast to large scales (>3 kpc) where dynamical perturbations like galaxy collisions, mergers or mass accretion as well as bars, spirals, and their gravity torques were identified as being responsible for the infall of gas, the effective process of removing angular momentum at small scales (100 pc–3 kpc) is not yet really revealed. Besides various scenarios invoking mini-bars, -spirals, warped nuclear disks, as well as lopsidedness or $m = 1$ instabilities, also the transport of the gas via hydronamically driven turbulences is a topic of recent discussions (Duschl et al., 2000).

Here, we will mainly concentrate on a brief overview about the results obtained for NGC3718 as well as HE1028, 3C48, and Q0957 + 561.

2. NGC 3718: A Nearby Warped Galaxy at $z = 0.03$

NGC3718 is an active galaxy of LINER 1.9 type (Ho et al., 1997) at a distance of 13Mpc ($z = 0.03$) belonging to the NUGA sample (García-Burillo et al., 2004). Its impressive dust and gas lane shows a warp signature extending over the entire bulge up to 1.5 arcmin away from the nucleus with a bending angle of almost 90° in the outer parts (Schwarz et al., 1985). The warped disk together with the close companion NGC3729 indicates a possible but not yet confirmed interaction of these two galaxies. The recently obtained PdBI mosaic map of the CO(1–0) emission contains different maxima supporting the warp model. Based on detailed simulations with a tilted ring model (Krips et al., in preparation; Pott et al., 2004), these different features (in total 5 labelled as E1, E2, C, W1, W2; see Figure 1) can be explained by projection, i.e. the so called orbit crowding effects. Assuming a standard $M_{\rm gas}(= M({\rm H_2} + {\rm He}) = 1.36 \times M({\rm H_2}))$ to $L'_{\rm CO(1-0)}$ ratio of $\sim 5\ \mathcal{M}_\odot$ (K kms pc^2)$^{-1}$ (Downes et al., 1993), we estimate the total molecular gas mass $M_{\rm gas}$ to $\sim 2.4 \times 10^8\ \mathcal{M}_\odot$. Compared to other galaxies from the NUGA sample, NGC3718 is hence rather poor in molecular gas (by an order of magnitude approximately). By zooming into the central region of NGC3718, the molecular

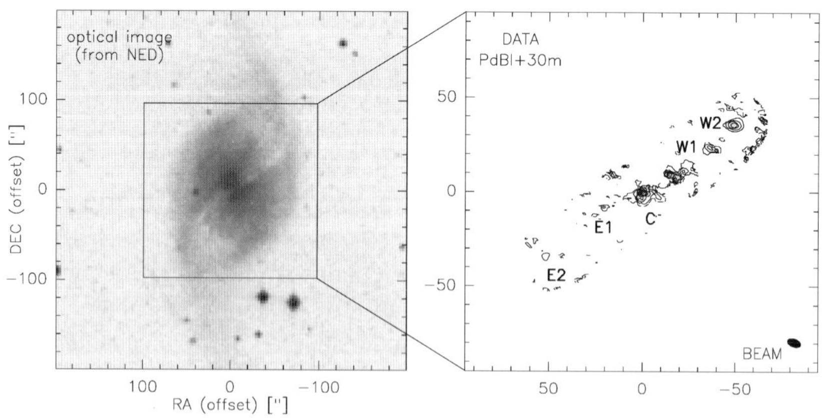

Figure 1. NGC3718 in the optical (*left*; from NED) and in CO(1–0) (*right*) as observed with the IRAM PdBI in mosaic mode. Contour levels: 1.0 to 5.0 Jy km/s by 0.5 Jy km/s. Positions with respect to: $\alpha_{\rm J2000} = 11{:}32{:}34.840$ $\delta_{\rm J2000} = +53{:}04{:}04.90$.

gas is not well centered on the radio/optical nucleus. However, we still find a weak velocity gradient across the nucleus indicating the kinematics of nuclear gas. A possible hypothesis how the gas can be transported to the inner parts might be given by Duschl et al. (2000) discussing the influence of hydrodynamically driven turbulences on gas accretion in disks.

3. HE1029: A Nearby QSO at $z = 0.04$

HE1029 is one of the closest known CO-bright QSOs hosts at a red-shift of $z = 0.04$. In the optical, it shows a strong bar in north-east to south-west direction. New PdBI observations of the CO(1–0) and CO(2–1) resulted in a clear detection of extended emission which appears to be correlated with the optical bar (Figure 2). Assuming a standard conversion factor (Downes et al., 1993), we find a molecular gas mass of $\sim 3 \times 10^9\ \mathcal{M}_\odot$. The lower limit on the dynamical mass of $\geq 9 \times 10^8\ \mathcal{M}_\odot$ together with the detected line profile suggests that the gas is seen rather face on. The line temperature ratio $R_{21} = \mathrm{CO(2\text{–}1)/CO(1\text{–}0)}$ of $\simeq 0.6$ implies subthermally excited, cold gas, similar to typical disk gas.

4. 3C48: A Merger at Intermediate Red-shift ($z = 0.37$)?

3C48 – at a red-shift of $z = 0.37$ – represents one of the first known quasar hosts. In the optical/infrared (Zuther et al., 2004) as well as in the radio domain two bright nuclear components are visible indicating a possible merger event already intensively discussed in the past. New PdBI observations of the CO(1–0) line which

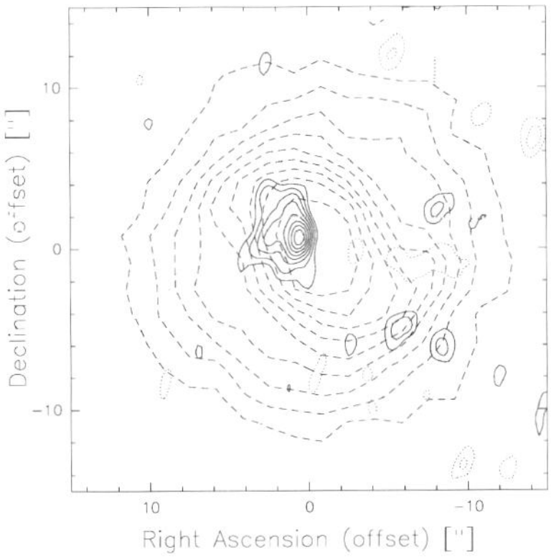

Figure 2. HE1029 in the optical (*dashed contours*; from NED) and in CO(2–1) (*solid contours*: ($2\sigma =$)1.2 to 5.4 Jy km/s by 0.6 Jy km/s) as observed with the IRAM PdBI. Positions with respect to: $\alpha_{\mathrm{J2000}} = $ 10:31:57.300 $\delta_{\mathrm{J2000}} = -$18:46:34.00.

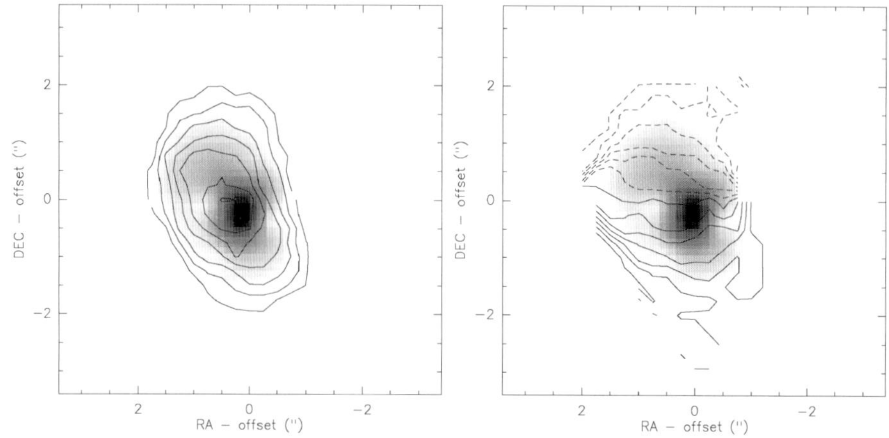

Figure 3. Overlay of an infrared map of 3C48(*grey scale;* Zuther et al. (2004)) with the integrated CO(1–0) emission (*left; black contours*: 0.2 to 0.8 Jy km/s by 0.1 Jy km/s) and with the velocity map of the CO(1–0) line (*right; black contours*: −95 to 35 km/s by 10 km/s) as measured with the IRAM PdBI. Positions with respect to: $\alpha_{J2000} = 01{:}37{:}41.300$ $\delta_{J2000} = +33{:}09{:}35.00$.

were combined with a previous data set by Wink et al. (1997) unveil gas rotation on the two "nuclear" components (Figure 3). This gas rotation is not in contradiction to recent multi-particle simulation of the stellar dynamics in its host galaxy (Scharwächter et al., 2004).

5. Q0957 + 561: Lensed CO Emission from a Disk at $z = 1.41$?

Q0957 + 561 was the first confirmed gravitationally lensed system (Walsh et al., 1979). Its lens consists of a giant elliptical galaxy at $z = 0.36$ with a surrounding cluster at the same red-shift and probably another group of background galaxies at $z = 0.5$ (Angonin-Willaime et al., 1994; Chartas et al., 1998). We present here new CO(2–1) observations carried out with PdBI. This data set was combined with a previous one by Planesas et al. (1999) to improve on sensitivity (Krips et al., 2003). A clear line profile is detected towards CO-A and CO-B (see Figure 4). However, the measured line profiles are quite different. Towards CO-A, we find two lines whereas in CO-B only a single blue-shifted line is visible. This difference can be explained by the gravitational lensing effect in this system (for more details see Krips et al., 2003, in press). The unlensed position of the quasar with respect to the lens is such as that the part closer to the lens (quasar itself with the blue shifted gas in the host galaxy) is still multiply lensed while the part further away (red-shifted region) is only magnified but not deflected into several images. Due to different arguments, we favour as hypothesis for the origin of this line profile a rotating gas disk in the host galaxy surrounding the quasar Q0957+561. One of the most important results is the detection of the host galaxy in the infrared/optical by

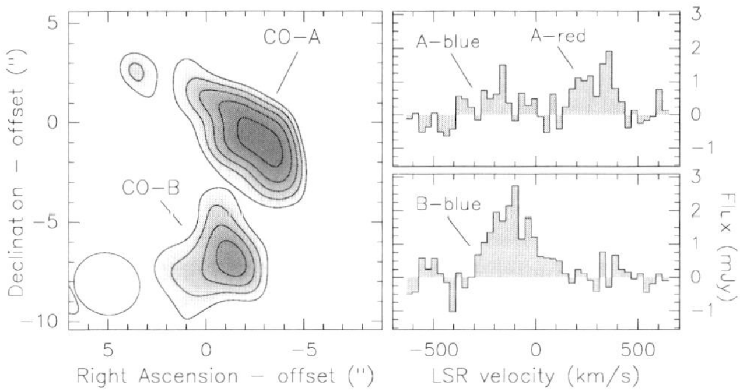

Figure 4. Integrated CO(2–1) emission of the lensed quasar Q0957+561 (*left*) observed with the PdBI. The spectrum of the CO(2–1) at CO-A and CO-B respectively are also shown (*right*). Contours: 0.33 to 77 Jy km/s by 0.11 Jy km/s. Positions with respect to: α_{J2000} = 10:01:21.07, δ_{J2000} = +55:53:56.7.

Keeton et al. (2000) agreeing with our results in shape and intensity ratio between the two lensed images of the host galaxy. The total molecular gas mass can be estimated to $5\times10^{10}\ \mathcal{M}_{\odot}$ assuming the same conversion factor as above. The upper limit on the line ratio R_{21} = CO(5–4)/CO(2–1) $\lesssim 1$ indicates low excitation conditions and agrees hence with global CO emission from a disk. Besides HR10 (Andreani et al., 2000), Q0957 + 561 is the only system where CO could be confirmed at such a red-shift range between 1 and 2!

6. Summary

We have observed the CO emission in five different active galaxies, four of which have been presented here with first results. Among them, three objects are classified as quasar/QSOs or at least quasar like (HE1029, 3C48 and Q0957 + 561) while the remaining two are rather low luminous active galaxies belonging to the so called group of LINERs. In all of them, CO gas could be detected. Assuming a standard conversion factor based on Downes et al. (1993), we find molecular gas masses from few times $10^8\ \mathcal{M}_{\odot}$ to $10^{10}\ \mathcal{M}_{\odot}$. The gas in all objects appear to be subthermally excited and cold characterizing low excitation conditions. This emphasizes the hypothesis that the gas in our objects is connected to a disk, even in the case of the higher red-shifted gravitationally lensed quasar Q0957 + 561.

Acknowledgement

Based on observations carried out with the IRAM PdBI. IRAM is supported by INSU/CNRS (france), MPG (Germany) and IGN (Spain). This work was partially financed by the SFB 494.

At present, it remains open whether such a minor merger accounts for the strong starburst and the nuclear activity of I Zw 1, as the possibility of a previous merger cannot be definitely excluded. At least a recent major merger seems unlikely regarding the strong impact of a major merger on morphology (e.g. Barnes and Hernquist, 1992, or the below-mentioned simulations for 3C 48).

3. Merger Properties of 3C 48

3C 48 ($z = 0.367$) resides in a clearly disturbed host galaxy (Figure 1, upper right panel). The host has a significant tidal tail to the north. Doubts about the merger scenario for 3C 48 stem from the apparently missing counter tidal tail (e.g. Boyce et al., 1999; Canalizo and Stockton, 2000) which is usually formed during the merger of two massive galaxies. The bright object to the south-east is either a background or a foreground galaxy (Canalizo and Stockton, 2000; Zuther et al., 2004). A second luminosity peak (3C 48A), detected $\sim$1 arcsec north-east of the 3C 48 quasar (Stockton and Ridgway, 1991; Zuther et al., 2004) could be the center of a merging companion galaxy. However, the radio-jet of the 3C 48 quasar points into the same direction so that a jet interaction with the interstellar medium is just as likely (Chatzichristou et al., 1999; Canalizo and Stockton, 2000).

The lower right panel of Figure 1 displays the model for 3C 48 based on the 3-dimensional, self-gravitating N-body simulations described in Scharwächter et al., 2004. In this model, the morphology and the stellar line-of-sight velocity field of 3C 48 are reproduced by the remnant of a major merger between two equal-mass disk galaxies, both inclined with respect to the orbital plane. The spatial structure of the remnant is complex and strongly depends on the viewing angle. For the particular tilt of the 3C 48 model, the counter tidal tail is located in front of the main galaxy and almost indistinguishable due to projection. As a by-product, this model allows for 3C 48A to be the center of the second merging galaxy.

4. Joint Discussion of I Zw 1 and 3C 48

Assuming an ongoing minor merger in the case of I Zw 1 and a nearly completed major merger for 3C 48, the following scenarios seem plausible: 3C 48 has the cumulative gas reservoir of two gas-rich precursor galaxies. The dynamical mass involved in the merger is large and the interaction is strong. Therefore, 3C 48 shows unusually large amounts of molecular gas in the center ($M_{H_2} = 2.7 \times 10^{10} M_{\odot}$, Wink et al., 1997). Merging with a low-mass elliptical dwarf, I Zw 1 is only supplied with its own original gas reservoir ($M_{H_2} \approx 9 \times 10^9 M_{\odot}$, Schinnerer et al., 1998) and the interaction is weak. However, the merging dwarf is able to cause dynamical instabilities in the I Zw 1 disk which stimulate gas inflow, starbursts, and nuclear activity. Since tidal forces are weak and the gas reservoir is small, I Zw 1 is only at the limiting luminosity between the ULIRG and QSO class ($L_{\rm IR} = 10^{11.87} L_{\odot}$, $M_{\rm B} = -22.62$, Canalizo and Stockton, 2001[1]), while 3C 48 is a more powerful counterpart ($L_{\rm IR} = 10^{12.81} L_{\odot}$, $M_{\rm B} = -24.55$, Canalizo and Stockton, 2001[1]). The starburst in I Zw 1 is a few 10 Myr old (Schinnerer et al., 1998). However, these young ages still allow for an already long-lasting interaction with the companion, considering that a typical merger happens on time scales of ~1 Gyr and that during an unequal-mass merger the starburst in the more massive galaxy can be delayed by up to 100 Myr compared to the starburst in the less-massive counterpart (Bernloehr, 1993). Although the minor merger in I Zw 1 is far from being complete in contrast to the major merger in 3C 48, the total time elapsed since the onset of the interactions could be similar in both cases. This admits the interpretation that the total interaction process between I Zw 1 and its companion is at a similar stage as the advanced major merger in 3C 48.

Regarding such scenarios, both QSOs seem to comply the parts they are assigned to as template objects at a later stage in the proposed evolutionary sequence.

Acknowledgements

We kindly thank L. Hernquist for providing the codes TREESPH and BUILDGAL and for giving helpful advice. This research is supported in part by the Deutsche Forschungsgemeinschaft (DFG) via grant SFB494. J. Scharwächter is supported by a scholarship for doctoral students of the Studienstiftung des deutschen Volkes.

Note

1. given for $H_0 = 75$ km s^{-1} Mpc^{-1} and $q_0 = 0.5$.

References

Barnes, J.E. and Hernquist, L: 1992, *ARA&A* **30**, 705.
Bernloehr, K.: 1993, *A&A* **268**, 25.
Boyce, P.J., Disney, M.J. and Bleaken, D.G.: 1999, *MNRAS* **302**, L39.
Bushouse, H.A., Borne, K.D., Colina, L., Lucas, R.A., Rowan-Robinson, M., Baker, A.C., Clements, D.L., Lawrence, A. and Oliver, S.: 2002, *ApJS* **138**, 1.
Canalizo, G. and Stockton, A.: 2000, *ApJ* **528**, 201.
Canalizo, G. and Stockton, A.: 2001, *ApJ* **555**, 719.
Chatzichristou, E.T., Vanderriest, C. and Jaffe, W.: 1999, *A&A* **343**, 407.
Hutchings, J.B. and Crampton, D.: 1990, *AJ* **99**, 37.
Lim, J. and Ho, P.T.P.: 1999, *ApJ* **510**, L7.
Sanders, D.B., Soifer, B.T., Elias, J.H., Madore, B.F., Matthews, K., Neugebauer, G. and Scoville, N.Z.: 1988, *ApJ* **325**, 74.

Scharwächter, J., Eckart, A., Pfalzner, S., Moultaka, J., Straubmeier, C. and Staguhn, J.G.: 2003a, *AaP* **405**, 959.
Scharwächter, J., Eckart, A., Pfalzner, S., Staguhn, J. and Schinnerer, E.: 2003b, *Ap & SS* **284**, 507.
Scharwächter, J., Eckart, A., Pfalzner, S., Zuther, J., Krips, M. and Straubmeier, C.: 2004, *A&A* **414**, 497.
Schinnerer, E., Eckart, A. and Tacconi, L. J.: 1998, *ApJ* **500**, 147.
Staguhn, J.G., Schinnerer, E., Eckart, E. and Scharwächter, J.: 2004, *ArXiv Astrophysics e-prints*.
Stockton, A. and Ridgway, S.E.: 1991, *AJ* **102**, 488.
Wink, J.E., Guilloteau, S. and Wilson, T.L.: 1997, *A&A* **322**, 427.
Zuther, J., Eckart, A., Scharwächter, J., Krips, M. and Straubmeier, C.: 2004, *A&A* **414**, 919.

H_2O MEGAMASERS: ACCRETION DISKS, JET INTERACTION, OUTFLOWS OR MASSIVE STAR FORMATION?

C. HENKEL[1], J.A. BRAATZ[2], A. TARCHI[3,4], A.B. PECK[5], N.M. NAGAR[6], L.J. GREENHILL[7], M. WANG[8] and Y. HAGIWARA[9]

[1]*Max-Planck-Institut für Radioastronomie, Auf dem Hügel 69, Bonn, Germany;*
E-mail: chenkel@mpifr-bonn.mpg.de
[2]*NRAO, Green Bank, WV, USA*
[3]*Istituto di Radioastronomia, CNR, Via Gobetti 101, Bologna, Italy*
[4]*INAF-Osservatorio Astronomico di Cagliari, Loc. Poggio dei Pini, Strada 54, Capoterra (CA), Italy*
[5]*Harvard-Smithsonian Center for Astrophysics, SAO/SMA Project, 645 N. A'ohoku Pl., Hilo, HI, USA*
[6]*Kapteyn Instituut, AV Groningen, The Netherlands*
[7]*Harvard-Smithsonian Center for Astrophysics, 60 Garden Street, Cambridge, MA, USA*
[8]*Purple Mountain Observatory, Chinese Academy of Sciences, Nanjing, China*
[9]*ASTRON/Westerbork Radio Observatory, NL-Dwingeloo 7990 AA, The Netherlands*

(Received 16 April 2004; accepted 15 June 2004)

Abstract. The 25 years following the serendipitous discovery of megamasers have seen tremendous progress in the study of luminous extragalactic H_2O emission. Single-dish monitoring and high-resolution interferometry have been used to identify sites of massive star formation, to study the interaction of nuclear jets with dense molecular gas and to investigate the circumnuclear environment of active galactic nuclei (AGN). Accretion disks with radii of 0.1–3 pc were mapped and masses of nuclear engines of order 10^6–10^8 $M_\odot$ were determined. So far, ~50 extragalactic H_2O maser sources have been detected, but few have been studied in detail.

Keywords: masers

1. Introduction

To date, maser lines of five molecular species, those of CH, OH, H_2O, SiO, and H_2CO, have been detected in extragalactic space. While the number of observed OH masers, ~100, is largest, the greatest emphasis of astrophysical research is focused on H_2O. This is a consequence of the fact that the 6_{16}–5_{23} line of interstellar water vapor at 22.235 GHz ($\lambda \sim 1.3$ cm), first detected toward Orion-KL, Sgr B2 and W 49 (Cheung et al., 1969) and, outside the Galaxy, toward the nearby spiral M 33 (Churchwell et al., 1977), can be efficiently used to pinpoint sites of massive star formation and to elucidate the properties of active galactic nuclei (AGN). The line traces dense ($n(H_2) \gtrsim 10^7$ cm^{-3}) warm ($T_{kin} \gtrsim 400$ K) molecular gas. Characteristic properties are enormous brightness temperatures (~10^{12} K has been measured), small sizes of individual hotspots ($\lesssim 10^{14}$ cm in galactic sources) and

Astrophysics and Space Science **295:** 107–116, 2004.

narrow linewidths (typically a few km s^{-1}) that make these masers to ideal probes of the structure and dynamics of the gas in which they reside. Because the apparent isotropic luminosity of these masers can be truly outstanding, reaching 10^{3-4} $L_\odot$ (up to 10^{53} photons/s) in the most extreme cases, H_2O masers can be observed out to fairly large distances (up to $cz \sim 17{,}700$ km s^{-1}; Tarchi et al., 2003).

2. Kilomasers Related to Star Formation

In the Galaxy, H_2O masers from star forming regions have isotropic luminosities of order $L_{H_2O} \lesssim 10^{-2}\,L_\odot$, with the notable exception of W 49 ($L_{H_2O} \sim 1\,L_\odot$). The more luminous of these sources consist of a number of hotspots (see Brand et al., this volume). In nearby galaxies, masers with similar luminosities, the so-called kilomasers, are also found with luminosities up to a few $L_\odot$. Such masers are important to pinpoint sites of massive star formation (e.g. Tarchi et al., 2002) and to estimate distances on a purely geometric basis, comparing radial velocity and proper motion dispersions in groups of maser spots (Greenhill et al., 1993; Argon et al., in press). Making use of known rotational properties, proper motions can also be used to determine three-dimensional velocity vectors of entire galaxies (e.g. Brunthaler et al., 2002), thus providing information on the content and distribution of mass inside the Local Group.

3. Accretion Disk Megamasers

Megamasers, with $L_{H_2O} \gtrsim 20\,L_\odot$, were first observed toward NGC 4945 (Dos Santos and Lépine, 1979) and then in NGC 1068, NGC 3079, NGC 4258 and the Circinus galaxy (Gardner and Whiteoak, 1982; Claussen et al., 1984; Henkel et al., 1984; Haschick and Baan, 1985). Two of these masers, those of NGC 1068 and NGC 4258, were soon found to be located in the innermost few parsecs of their parent galaxies (Claussen and Lo, 1986). Selection criteria to find more such masers as well as an interpretation of the molecular line emission in terms of properties characterizing the nuclear environment of AGN remained, however, elusive for a full decade. A large survey including ~360 Seyfert and LINER galaxies (Braatz et al., 1996) finally led to the identification of another 10 megamasers, thus tripling the number of known sources and permitting, for the first time, a statistical analysis. All of these masers were found to be associated with Seyfert 2 or LINER nuclei (Braatz et al., 1997). Adopting the so-called unified scheme in which Seyfert 1s and 2s are identical except for angle of view, H_2O megamaser activity then is related to the large line-of-sight column densities expected when the nuclear tori are viewed edge-on. At least some LINER galaxies also contain large column densities of warm dense molecular gas as well as an AGN.

More detailed studies of the megamaser in NGC 4258 not only confirmed the presence of >10 systemic velocity features at a given time but also led to the detection of two additional H_2O velocity components, $\pm 1000\,\mathrm{km\,s^{-1}}$ off the systemic velocity (Nakai et al., 1993). The systemic features show a secular drift, with radial velocities increasing by $dV_s/dt \sim 10\,\mathrm{km\,s^{-1}\,yr^{-1}}$ (Haschick and Baan, 1990; Haschick et al., 1994; Greenhill et al., 1995a; Nakai et al., 1995). Very long baseline interferometry (VLBI) reveals the presence of a warped edge-on Keplerian disk of $\sim$0.5 pc diameter (Greenhill et al., 1995b; Miyoshi et al., 1995; Herrnstein et al., 1998a, Herrnstein et al., 1999), as shown in Figure 1. Its geometry reflects the velocity coherence of the differentially rotating gas along the lines-of-sight toward its front side, its back side and its tangentially viewed regions (Figure 2) as well as possible amplification of the radio continuum of the northern jet by the systemic features.

The positive velocity drift of the systemic component is readily explained by centripetal acceleration, implying that the features arise from the front and not from the back side of the masing disk (the latter would result in a negative drift). The 'high velocity features' originate from those parts of the disk seen tangentially. Applying the virial theorem yields

$$M_{\mathrm{core}} = 1.12 \left[\frac{V_{\mathrm{rot}}}{\mathrm{km\,s^{-1}}}\right]^2 \left[\frac{R}{\mathrm{mas}}\right] \left[\frac{D}{\mathrm{Mpc}}\right] M_{\odot}, \tag{1}$$

with M_{core} being the mass enclosed by the Keplerian disk, V_{rot} denoting its rotational velocity at angular radius R, and D representing the distance to the galaxy. The VLBI maps allow us to directly measure V_{rot} and R for various values of R. From the Keplerian rotation curve we then obtain

$$C_1 = \left[\frac{V_{\mathrm{rot}}}{\mathrm{km\,s^{-1}}}\right] \left[\frac{R}{\mathrm{mas}}\right]^{1/2}. \tag{2}$$

In addition, the observed E-W velocity gradient of the systemic features (Figure 1c) provides

$$C_2 = \left[\frac{V_{\mathrm{rot}}}{\mathrm{km\,s^{-1}}}\right] \left[\frac{R}{\mathrm{mas}}\right]^{-1}. \tag{3}$$

$C_1/C_2 = R^{3/2}$ then gives the angular radius R_s of the systemic features as viewed from a direction in the plane of the disk, but perpendicular to the line-of-sight. The result, $R_s \sim 4.1$ mas, implies that the features are localized toward the inner edge of the disk. The distance to the disk, needed to estimate M_{core} in Eq. (1), is determined by measuring the centripetal acceleration

$$dV_s/dt = \frac{V^2_{\mathrm{rot,s}}}{r_s} = 9.3 \pm 0.3\,\mathrm{km\,s^{-1}\,yr^{-1}}. \tag{4}$$

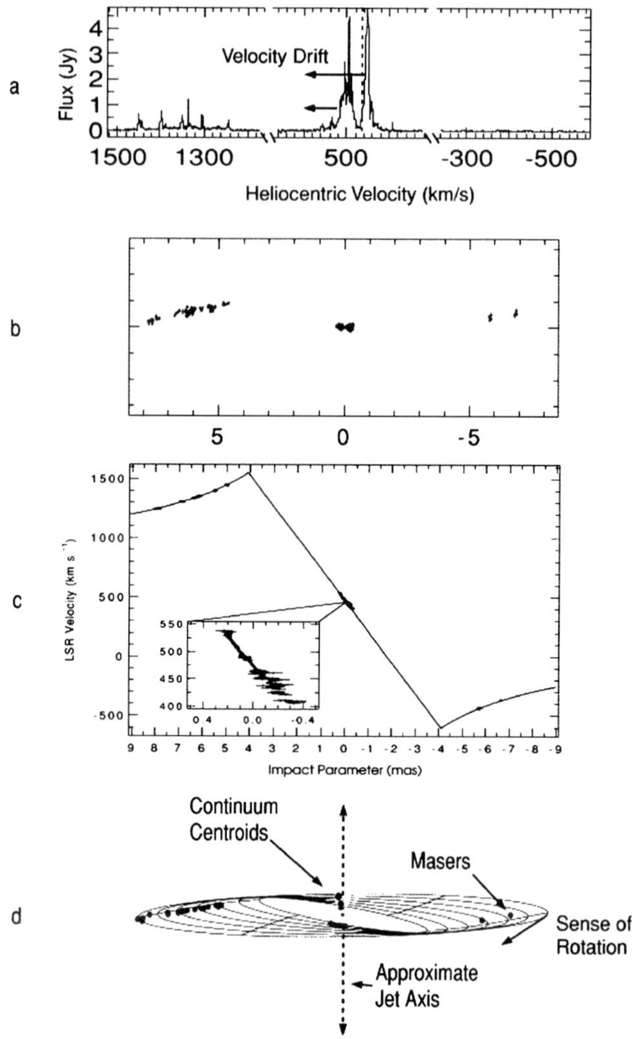

Figure 1. Overview of the NGC 4258 system: (a) typical spectrum, (b) very long baseline array (VLBA) map, (c) velocity versus impact parameter with Keplerian rotation curve fit, and (d) warped disk with VLBA map (from Bragg et al., 2000).

With dV_s/dt and the rotational velocity $V_{rot,s}$ of the systemic components known, the linear scale r_s can be compared with the angular scale, R_s, providing a measure of the distance. The proper motion of the individual systemic maser spots, $31.5 \pm 1.0\ \mu$as yr^{-1}, allows, with $V_{rot,s}$ being known, a check of this distance estimate. The results of the two methods agree. $D = 7.2 \pm 0.5$ Mpc and $M_{core} = (3.9 \pm 0.3) \times 10^7\ M_\odot$ within 0.14 pc (Herrnstein et al., 1999).

So far, all H_2O megamasers studied interferometrically arise from the innermost parsecs of their parent galaxy and all appear to be associated with AGN. Searching

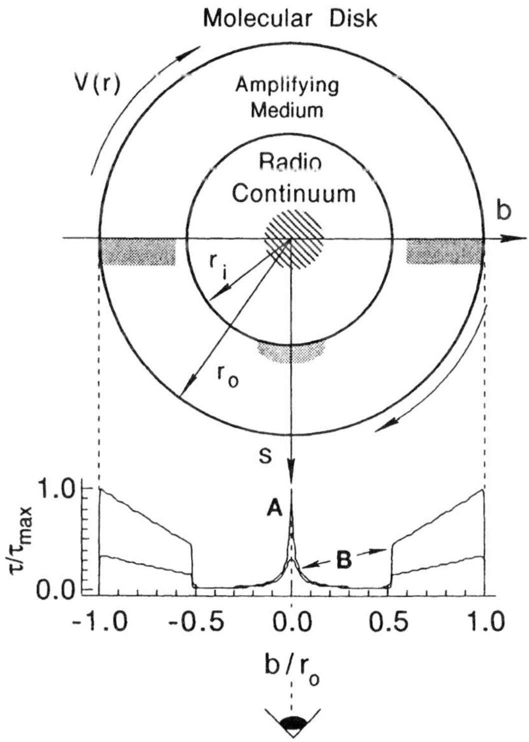

Figure 2. The Keplerian disk of NGC 4258 seen from above. The normalized optical depth along various lines of sight is shown for the plane of the disk, where b is the impact parameter. The shaded areas denote lines-of-sight with detected maser emission. $r_{\rm i}$ and $r_{\rm o}$ are the inner and outer radii of the masing disk, s marks the distance from its midline (from Greenhill et al., 1995a).

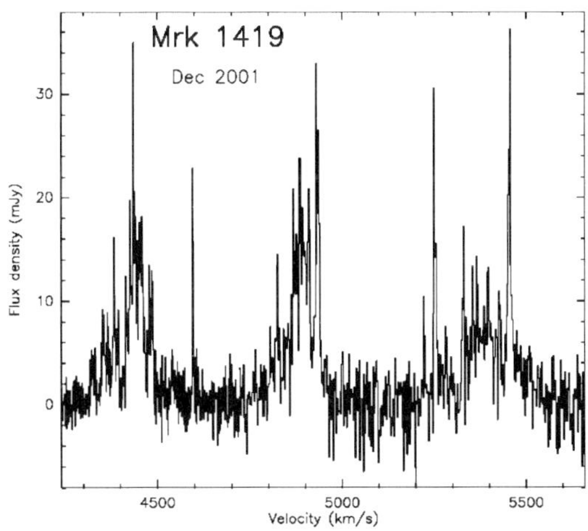

Figure 3. 22 GHz H_2O spectrum of Mrk 1419 (NGC 2960), taken with the 100-m telescope at Effelsberg (from Henkel et al., 2002) and showing the characteristic systemic (center), red- and blue-shifted groups of maser features.

for NGC 4258-like targets showing three groups of H_2O features, one systemic, one red- and one blue-shifted, has become the Holy Grail of recent maser surveys. Such a configuration requires not only the presence of very dense molecular gas, but also a suitable viewpoint, with the nuclear disk seen almost edge-on. Therefore such sources are rarely found. Nevertheless, the number of detected sources falling into this category is steadily rising (see e.g. Figure 3). Most notable are IC 2560 (2900 km s^{-1}; Ishihara et al., 2001) and Mrk 1419 (4900 km s^{-1}; Henkel et al., 2002). Very recently, four additional sources were identified, one at 7900 km s^{-1} (Braatz et al., 2004) and three at >10,000 km s^{-1} (Greenhill et al., in preparation). Assuming that the model valid for NGC 4258 is generally applicable, single dish observations determining the drift of the systemic components (dV_s/dt) and the velocity offsets to the non-systemic components (V_{rot}) are sufficient for a rough estimate of the enclosed mass and linear scale of the disk. Thus it is possible to estimate angular scales of order 1 mas observing with resolutions of 35–40″. So far, all measured centripetal accelerations are positive. The most plausible explanations are obscuration of the backside, perhaps by free–free absorption, and a lack of background radio continuum emission that could be amplified.

Maser dynamical masses (M_{core}) for a large sample of galaxies have the potential to establish the slope of the M_{core}–σ relation and its intrinsic scatter (e.g. Ferrarese and Merritt, 2000), with uncertainties being dominated by the stellar velocity dispersion σ. Observing the three-dimensional structure of a sample of accretion disks may provide strong constraints on heating (possibly irradiation by the nuclear X-ray source) and warping mechanisms (radiative torques have been proposed) and on the stability of these structures (e.g. Neufeld and Maloney, 1995). The thickness of the respective accretion disk is another important parameter (e.g. Moran et al., 1999), being crucial to calculations of accretion rate and identification of accretion modes (e.g. advective, convective, viscous). Determining the mass of the nuclear source, its distance, Eddington luminosity and accretion efficiency, the rate and mode of the nuclear accretion flow, the size and geometry of the nuclear disk or torus, the parent galaxy's deviation from the Hubble flow and calibrations of optical or near infrared indicators of distance with geometrically obtained values (see e.g. Eqs. (1)–(4)) are all attractive goals.

In addition to the sources introduced above, there are a large number of targets exhibiting somewhat less regular spectra. These may either arise from tori that are unstable to fragmentation and star formation or from sources that combine an edge-on nuclear accretion disk with maser components of different nature that will be discussed later. Within this context, the most thoroughly studied sources are NGC 1068, NGC 3079 and the Circinus galaxy (Gallimore et al., 1996; Greenhill et al., 1996, 2003; Greenhill and Gwinn, 1997; Trotter et al., 1998; Hagiwara et al., 2002; Kondratko et al., in press).

4. Jet Megamasers

There are also sources in which at least a part of the H_2O emission is believed to be the result of an interaction between the nuclear radio jet and an encroaching molecular cloud. The first such source was NGC 1068, where not only the three groups of H_2O components from the accretion disk are seen (Greenhill et al., 1996), but where a fourth component is also detected. This originates from a region $0''.3$ ($\sim$30 pc) downstream, where the radio jet bends (Gallimore et al., 1996, 2001). To maximize detection rates of jet-maser sources, galaxies should be selected that either show evidence for interaction between the radio jet and clouds in the narrow line region or that show a face-on ($i < 35^\circ$) large-scale galaxy disk and extended (up to 100 pc) radio structures, indicating that both the disk and the radio jet are fairly close to the plane of the sky.

Toward the elliptical galaxy NGC 1052 the H_2O maser features are located along the line-of-sight to the south-western nuclear jet (Claussen et al., 1998). In Mrk 348, a spiral galaxy with a particularly strong nuclear radio component, the megamaser appears to be associated with the northern jet (see Figure 4; Peck et al., 2003). The intensity of the line emission is correlated with the continuum flux. The high linewidth ($\sim$130 km s^{-1}) on small spatial scales ($<$0.25 pc) and the rapid variability indicate that the H_2O emission is more likely to arise from a shocked region at the interface between the energetic jet and the ambient molecular gas than as a result of amplification of the continuum jet by molecular clouds along the line-of-sight. The line emission, red-shifted by 130 km s^{-1} with respect to the systemic velocity, may arise from gas being entrained by the receding jet. The close temporal correlation between the flaring activity of the maser and the continuum further suggests that the H_2O and continuum hotspots are nearly equidistant from the central engine and may be different manifestations of the same event.

5. New Developments

Recently, optically detected large scale outflows were proposed to induce shocks that might trigger H_2O megamaser emission (Schulz and Henkel, 2003). Independently, it was found that in the Circinus galaxy the H_2O megamaser does not only trace a circumnuclear disk, but that there are also features associated with a wide angle outflow (Greenhill et al., 2003). H_2O maser emission traces this outflow out to 1 pc from the central engine with line-of-sight velocities up to $\pm$160 km s^{-1} w.r.t. systemic. The outflowing wind is observed in those regions around the circumnuclear disk that are not shadowed by its warp. The position angles of the edges of the outflow correspond to those of the outflow and ionization cones observed at radio and optical wavelengths on much larger scales (Veilleux and Bland-Hawthorn, 1997; Curran et al., 1999). Thus in addition to accretion disk and jet megamasers, strong maser emission can also be associated with large scale

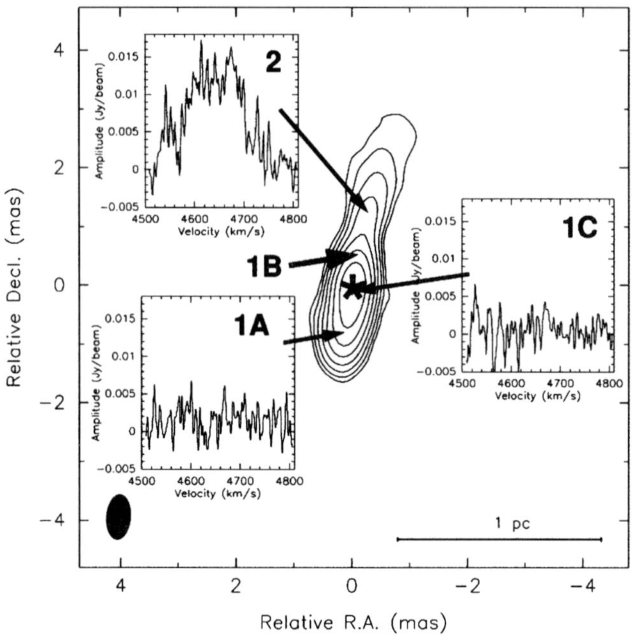

Figure 4. VLBA 22 GHz continuum and H_2O spectra toward Mrk 348 (NGC 262; see Peck et al., 2003). H_2O emission is only detected toward the northern jet.

nuclear outflows. Yet another kind of H_2O megamaser might be associated with massive star formation. So far there is not yet a confirmed source. Toward Arp 299, an interacting pair of galaxies, luminous H_2O megamaser emission was detected that might be associated with star formation in the overlapping region between its main members NGC 3690 and IC 694 (Henkel et al., in press). This possibility has, however, to be confirmed by interferometric measurements.

As suggested by Section 2, most H_2O kilomasers are associated with star formation. There are, however, exceptions. Toward M 51, the kilomaser was found to coincide with the nucleus within 250 mas (~10 pc; Hagiwara et al., 2001). A similarly good coincidence, within 1″ (~12 pc), was also found for the main maser component in the starburst galaxy NGC 253 (Henkel et al., 2004). The nature of these sources that are too weak to be seen at distances well in excess of 10 Mpc is still under discussion.

Almost a decade ago, all H_2O megamasers were found to be associated with Seyfert 2 and LINER galaxies (Braatz et al., 1997). In the meantime, however, with a much larger number of detections, the situation has become more complex. While megamasers are detected in optically 'normal' galaxies (Greenhill et al., 2002), a particularly luminous megamaser was recently found in the FRII radio galaxy 3C 403 (Tarchi et al., 2003). Another megamaser host, NGC 5506, was identified as a Narrow Line Seyfert 1 (Nagar et al., 2002). The detection of a nuclear kilomaser in a second such galaxy, NGC 4051 (Hagiwara et al., 2003), was interpreted in terms of a relatively low inclination of the nuclear disk or torus w.r.t Seyfert 2 galaxies, thus yielding relatively low column densities for coherent amplification. Even more recently, H_2O was detected in the prototypical Seyfert 1.5 NGC 4151 (Braatz et al., 2004). It is still too early for a detailed interpretation, but it is clear that these detections will be essential for a better understanding of the unifying scheme differentiating between type 1 and type 2 AGN.

Is there a way to predict the presence of H_2O megamasers by observations at other wavelengths? This question was recently discussed for the prototypical sources NGC 1068, NGC 3079 and NGC 4258 (Bennert et al., 2004). All three galaxies exhibit a spatially compact ($\lesssim 1''$) near infrared core containing dust clouds that are heated by the central engine. This appears to be the main hint for the potential presence of accretion disk masers, but these cores are more difficult to detect than the masers themselves. Tracers for jet masers at optical and near infrared wavelengths are spectral lines that are split into two velocity components.

6. Future Prospects

Though the sensitivity of existing facilities can be slightly enhanced, order(s) of magnitude improvements will only be possible with the square kilometer array (SKA), complemented by space interferometry to combine optimal sensitivity with optimal angular resolution. The 22 GHz line of water vapor bears the prospect to reveal the magnetic field strength through its Zeeman pattern. A measurement toward NGC 4258 yielded a 1σ upper limit of 300 mG for the toroidal component (Herrnstein et al., 1998b). Probably only the SKA will provide the sensitivity to determine the magnetic field in the circumnuclear environment of active galaxies. The same may hold for searches of H_2O megamasers at cosmological red shifts. On the other hand, accounting for the statistical properties of detected maser sources, it appears that existing facilities have the potential to drastically enlarge the number of known luminous sources of H_2O emission. So far, possibly only a small percentage of the detectable megamasers have been found (Peck et al., in press).

References

Argon, A.L., Greenhill, L.J., Reid, M.J. et al.: in press, *ApJ*.
Bennert, N., Schulz, H. and Henkel, C.: 2004, *A&A* **419**, 127.
Braatz, J.A., Wilson, A.S. and Henkel, C.: 1996, *ApJS* **106**, 51.
Braatz, J.A., Wilson, A.S. and Henkel, C.: 1997, *ApJS* **110**, 321.
Braatz, J.A., Henkel, C., Greenhill, L.J., Moran, J.M. and Wilson, A.S.: 2004, *ApJ*, in preparation.
Bragg, A.E., Greenhill, L.J., Moran, J.M. and Henkel, C.: 2000, *ApJ* **535**, 73.
Brunthaler, A., Falcke, H., Reid, M., Greenhill, L.J. and Henkel, C.: 2002, in: Ros et al. (eds.), *Proceedings of the Sixth European VLBI Symposium*, MPIfR, Bonn, Germany, p. 189.
Cheung, A.C., Rank, D.M., Townes, C.H., Thornton, D.D. and Welch, W.J.: 1969, *Nature* **221**, 626.
Churchwell, E., Witzel, A., Huchtmeier, W. et al.: 1977, *A&A* **54**, 969.

Claussen, M.J., Heiligman, G.M. and Lo, K.-Y.: 1984, *Nature* **310**, 298.
Claussen, M.J. and Lo, K.-Y.: 1986, *ApJ* **308**, 592.
Claussen, M.J., Diamond, P.J., Braatz, J.A., Wilson, A.S. and Henkel, C.: 1998, *ApJ* **500**, L129.
Curran, S.J., Rydbeck, G., Johansson, L.E.B. and Booth, R.S.: 1999, *A&A* **344**, 767.
Dos Santos, P.M. and Lépine, J.R.D.: 1979, *Nature* **278**, 34.
Ferrarese, L. and Merritt, D.: 2000, *ApJ* **539**, L9.
Gardner, F.F. and Whiteoak, J.B.: 1982, *MNRAS* **201**, 13.
Gallimore, J.F., Baum, S.A., O'Dea, C.P., Brinks, E. and Pedlar, A.: 1996, *ApJ* **462**, 740.
Gallimore, J.F., Henkel, C., Baum, S.A. et al.: 2001, *ApJ* **556**, 694.
Greenhill, L.J. and Gwinn, C.R.: 1997, *ApSS* **248**, 261.
Greenhill, L.J., Moran, J.M., Reid, M.J., Menten, K.M. and Hirabayashi, H.: 1993, *ApJ* **406**, 482.
Greenhill, L.J., Henkel, C., Becker, R., Wilson, T.L. and Wouterloot, J.G.A.: 1995a, *A&A* **304**, 21.
Greenhill, L.J., Jiang, D.R., Moran, J.M. et al.: 1995b, *ApJ* **440**, 619.
Greenhill, L.J., Gwinn, C.R., Antonucci, R. and Barvainis, R.: 1996, *ApJ* **472**, L21.
Greenhill, L.J., Ellingsen, S.P., Norris, R.P. et al.: 2002, *ApJ* **565**, 836.
Greenhill, L.J., Booth, R.S., Ellingsen, S.P. et al.: 2003, *ApJ* **590**, 162.
Hagiwara, Y., Henkel, C., Menten, K.M. and Nakai, N.: 2001, *ApJ* **560**, L37.
Hagiwara, Y., Henkel, C., Sherwood, W.A. and Baan, W.A.: 2002, *A&A* **387**, L29.
Hagiwara, Y., Diamond, P.J., Miyoshi, M., Rovilos, E. and Baan, W.A.: 2003, *MNRAS* **344**, L53.
Haschick, A.D. and Baan, W.A.: 1985, *Nature* **314**, 144.
Haschick, A.D. and Baan, W.A.: 1990, *ApJ* **355**, L23.
Haschick, A.D., Baan, W.A. and Peng, E.W.: 1994, *ApJ* **437**, L35.
Henkel, C., Güsten, R., Downes, D. et al.: 1984, *A&A* **141**, L1.
Henkel, C., Braatz, J.A., Greenhill, L.J. and Wilson, A.S.: 2002, *A&A* **394**, L23.
Henkel, C., Tarchi, A., Menten, K.M. and Peck, A.B.: 2004, *A&A* **414**, 117.
Herrnstein, J.R., Greenhill, L.J., Moran, J.M. et al.: 1998a, *ApJ* **497**, L69.
Herrnstein, J.R., Moran, J.M., Greenhill, L.J., Blackman, E.G. and Diamond, P.J.: 1998b, *ApJ* **508**, 243.
Herrnstein, J.R., Moran, J.M., Greenhill, L.J. et al.: 1999, *Nature* **400**, 539.
Ishihara, Y., Nakai, N., Iyomoto, N. et al.: 2001, *PASJ* **53**, 215.
Kondratko, P.T., Greenhill, L.J. and Moran, J.M.: in press, *ApJ*.
Miyoshi, M., Moran, J.M., Herrnstein, J.R. et al.: 1995, *Nature* **373**, 127.
Moran, J.M., Greenhill, L.J. and Herrnstein, J.R.: 1999, *JApA* **20**, 165.
Nagar, N.M., Oliva, E., Marconi, A. and Maiolino, R.: 2002, *A&A* **391**, L21.
Nakai, N., Inoue, M. and Miyoshi, M.: 1993, *Nature* **361**, 45.
Nakai, N., Inoue, M., Miyazawa, K., Miyoshi, M. and Hall, P.: 1995, *PASJ* **47**, 771.
Neufeld, D.A. and Maloney, P.R.: 1995, *ApJ* **447**, L17.
Peck, A.B., Henkel, C., Ulvestad, J.S. et al.: 2003, *ApJ* **590**, 149.
Peck, A.B., Tarchi, A., Henkel, C. et al.: in press, *ApJ*.
Schulz, H. and Henkel, C.: 2003, *A&A* **400**, 41.
Tarchi, A., Henkel, C., Peck, A.B. and Menten, K.M.: 2002, *A&A* **389**, L39.
Tarchi, A., Henkel, C., Chiaberge, M. and Menten, K.M.: 2003, *A&A* **407**, L33.
Trotter, A.S., Greenhill, L.J., Moran, J.M. et al.: 1998, *ApJ* **495**, 740.
Veilleux, S. and Bland-Hawthorn, J.: 1997, *ApJ* **479**, L105.

PROBING THE OBSCURING MEDIUM AROUND ACTIVE NUCLEI USING MASERS: THE CASE OF 3C 403

A. TARCHI[1,2], C. HENKEL[3], M. CHIABERGE[1], K. M. MENTEN[3], A. BRUNTHALER[4] and L. MOSCADELLI[2]

[1]*Istituto di Radioastronomia, CNR, Via Gobetti 101, Bologna, Italy;*
E-mail: a.tarchi@ira.cnr.it
[2]*INAF-Osservatorio Astronomico di Cagliari, Loc. Poggio dei Pini, Strada 54, Capoterra (CA), Italy*
[3]*Max-Planck-Institut für Radioastronomie, Auf dem Hügel 69, Bonn, Germany*
[4]*JIVE, Joint Institute for VLBI in Europe, AA Dwingeloo, The Netherlands*

(Received 16 April 2004; accepted 15 June 2004)

Abstract. We report the first detection of a water megamaser in a radio-loud galaxy, 3C 403, and present a follow-up study using the VLA. 3C 403 has been observed as a part of a small sample of FR II galaxies with evidence of nuclear obscuration. The isotropic luminosity of the maser is $\sim$1200 $L_{\odot}$. With a recessional velocity of $cz \sim 17680$ km s^{-1} it is the most distant water maser so far reported. The line arises from the densest ($>10^8$ cm^{-3}) interstellar gas component ever observed in a radio-loud galaxy. Two spectral features are identified, likely bracketing the systemic velocity of the galaxy. Our interferometric data clearly indicate that these arise from a location within $0.1''$ ($\approx$110 pc) from the active galactic nucleus. We conclude that the maser spots are most likely associated with the tangentially seen parts of a nuclear accretion disk, while an association with dense warm gas interacting with the radio jets cannot yet be ruled out entirely.

Keywords: 3C 403, active galaxies, masers

1. Introduction

So far, water megamasers have been detected in radio-quiet active galactic nuclei (AGN; for a definition of radio-quiet AGN, see e.g. Kellermann et al., 1989), mostly in Seyfert 2 and LINER galaxies. Under the assumption that all megamasers are associated with molecular material orbiting around the central engine (e.g. Miyoshi et al., 1995) or interacting with the nuclear jet(s) of the host galaxy (e.g. Claussen et al., 1998) and that their amplification is unsaturated (i.e. the maser intensity grows linearly with the background radio continuum), one should expect a much higher detection rate in radio-loud AGN than is actually observed. This fact should be especially true for the radio-loud AGN classified as narrow-lined FR IIs for which the AGN unification scheme (e.g. Urry and Padovani, 1995) requires the presence of geometrically and optically thick obscuring structure.

Samples of radio galaxies belonging to the broad-lined and narrow-lined FR II class have been recently observed with the Effelsberg telescope (Tarchi et al., in

Astrophysics and Space Science **295:** 117–123, 2004.

(asymmetrically) bracketing the nominal systemic velocity of the galaxy ($cz = 17688$ km s^{-1}, see next section): the stronger one has a velocity of $cz = 17827 \pm 1$ km s^{-1}, a width of 31 ± 2 km s^{-1}, and a flux density peak of 23 ± 3 mJy; the weaker one has a velocity of $cz = 17644 \pm 5$ km s^{-1}, a width of 53 ± 8 km s^{-1}, and a peak flux density of 4.0 ± 0.5 mJy. Using a distance of 235 Mpc ($H_0 = 75$ km s^{-1} Mpc^{-1}), the total isotropic luminosities are 950 ± 140 and 280 ± 55 $L_{\odot}$ for the main components, respectively. Hence, the maser in 3C 403 is the most distant and one of the most luminous water maser so far reported (the most luminous maser, with 6000 $L_{\odot}$, and, before this discovery also the most distant one, is in TXFS2226-184; Koekemoer et al., 1995).

The outcome of the VLA observations (Figure 3; right-hand side) confirms the presence of the two main maser features, and indicates that the maser emission is unresolved and arise from a location within 0.1″ (≈110 pc) from the active galactic nucleus. The strength of the weaker line is consistent with the single-dish result, while the stronger line is much weaker than that obtained with Effelsberg. Because of the unresolved nature of the emission it is unlikely that the loss of flux density in the feature is due to extended emission resolved-out because of the higher resolution. More probably, the cause is a flare-down of the component due to strong variability, a phenomenon already well known to exist in water masers.

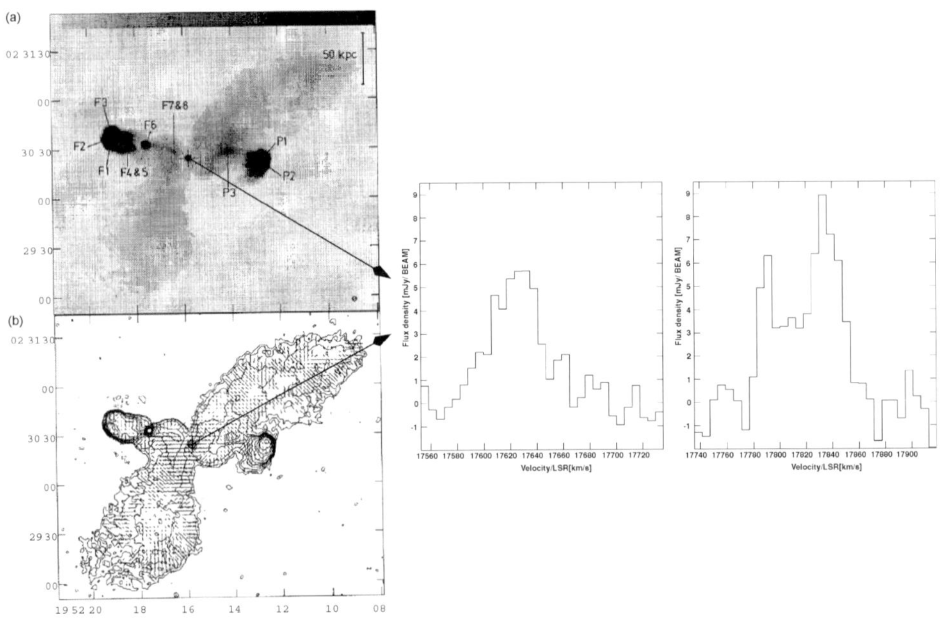

Figure 3. Left-hand side VLA 8.4 GHz maps in grey-scale (top) and contours (bottom) of 3C 403 at a resolution of ~2.5″ (FWHM). For details, see Black et al. (1992). *Right-hand side* the two main water maser features in 3C 403 observed with the VLA A-array. Both lines arise from a location within 0.1″ (≈110 pc) from the active galactic nucleus ($RA_{B1950} = 19^h\ 52^m\ 15.8^s$; $Dec_{B1950} = 02^{\circ}30'24''$).

5. Discussion and Conclusions

The standard unified scheme of AGN requires an obscuring region, possibly containing molecular gas that surrounds the central engine and that effectively shields the inner few parsecs from view, if the radio axis lies close to the plane of the sky (Antonucci, 1993). In the innermost part, at radii up to some tenths of a parsec, this material is likely to form a rapidly rotating accretion disk around a central supermassive black hole. At larger distances (up to about 50–100 pc) the atomic and molecular gas is possibly distributed in a toroidal structure providing obscuration of the central regions to particular lines-of-sight.

Indeed, very few direct detections of molecular gas in radio-loud galaxies have been reported so far (H_2 in Cygnus A: Evans et al., 1999; CO in 3C 293: Wilman et al., 2000). The detection discussed here strongly favors the presence of molecular material near the central engines of at least some FR IIs. Past negative results in molecular line surveys may also be a consequence of observational sensitivity limits (such a possibility was also mentioned by Henkel et al., 1998). Furthermore, because of the high gas densities required for H_2O masers to operate ($>10^8$ cm^{-3}; e.g. Elitzur et al., 1989), our detection represents the densest interstellar gas component ever observed in a radio-loud galaxy (typical values for molecular gas densities range between $\sim 10^3$ and $\sim 10^5$ cm^{-3}; e.g. Wilman et al., 2000).

5.1. Accretion Disk or Jet Interaction?

Because of the large uncertainty of the systemic velocity[3] (V_{sys}) of 3C 403, the following discussion and (preliminary) conclusions are based on the assumption that V_{sys} is placed (as indicated in Figure 2) between the two main H_2O maser components.

As mentioned in Section 1, interferometric studies of H_2O megamasers have shown that the emission is either associated with a nuclear accretion disk (for NGC 4258, see e.g. Miyoshi et al., 1995) or with the radio-jets interacting with dense molecular material near the center (for Mrk 348, see Peck et al., 2003).

The interferometric observation performed with the VLA in its A configuration confirms that also the megamaser in 3C 403 has a nuclear origin.

Only VLBI interferometric observations at milliarcsecond resolution (spatial scales of ≈ 1 parsec) will allow us to determine (or to provide an upper limit to) the extent of the emission and to pinpoint the exact location of the H_2O emitting region(s) in order to see if the maser is associated with an accretion disk or radio jets. Nevertheless, a qualitative discussion is possible on the basis of the single-dish spectra of Figure 2 and the new VLA observation.

Accretion disk: this scenario is particularly supported by the expected almost edge-on orientation of the nuclear obscuring layer (see Chiaberge et al., 2002). As in the case of the Seyfert galaxy NGC 4258 (e.g. Miyoshi et al., 1995), the spectrum should then show three distinct groups of features: one centered at the systemic velocity of the galaxy (originating along the line of sight to the nucleus) and two groups symmetrically offset from the systemic velocity, arising from those parts of the disk that are viewed tangentially. If the latter are the two lines we are observing, the rotational velocity of the disk is $\sim$100 km s^{-1} (which is, we have to point out, very small when compared with that derived for NGC 4258). In 3C 403 the systemic lines seem instead to be missing. To explain this fact, partly following the hypothesis proposed by Henkel et al. (1998), we could argue that the circumnuclear disk is actually a *thin rotating ring* of the type modeled by Ponomarev et al. (1994; their Figure 2), where masing gives a double-peaked profile only.

Jet interaction: the handful of bright knots visible in the radio images shown in Figure 3 (left-hand side) hints at the presence of jets interacting in several regions with the interstellar medium. The profile of the spectrum does not contradict a 'jet-origin' of the detected maser emission in 3C 403. If we assume a symmetric molecular distribution, the two observed features could be interpreted as the red-shifted and blue-shifted counterparts of the maser line, arising from opposite jets close to the core.

Notes

1. The 100-m telescope at Effelsberg is operated by the Max-Planck-Institut für Radioastronomie (MPIfR) on behalf of the Max-Planck-Gesellschaft (MPG).
2. The National Radio Astronomy Observatory is a facility of the National Science Foundation operated under cooperative agreement by Associated Universities, Inc.
3. Velocities derived from optical emission lines may be uncertain or biased by motions of the emitting gas (e.g. Morganti et al., 2001). From the rotation curve measured by Baum et al. (1990), we deduce that these uncertainties in 3C 403 are $<$ 100 km s^{-1}.

References

Antonucci, R.: 1993, *ARA&A* **31**, 473.
Baum, S.A., Heckman, T. and van Breugel, W.: 1990, *ApJS* **74**, 389.
Black, A.R.S., Baum, S.A., Leahy, J.P., et al.: 1992, *MNRAS* **256**, 186.
Chiaberge, M., Capetti, A. and Celotti, A.: 2002, *A&A* **394**, 791.
Claussen, M.J., Diamond, P.J., Braatz, J.A., Wilson, A.S. and Henkel, C.: 1998, *ApJ* **500**, L129.
Elitzur, M., Hollenbach, D.J. and McKee, C.F.: 1989, *ApJ* **346**, 983.
Evans, A.S., Sanders, D.B., Surace, J.A. and Mazzarella, J.M.: 1999, *ApJ* **511**, 730.
Gallimore, J.F., Henkel, C., Baum, S.A., et al.: 2001, *ApJ* **556**, 694.
Henkel, C., Wang, Y.P., Falcke, H., Wilson, A.S. and Braatz, J.A.: 1998, *A&A* **335**, 463.
Jackson, N. and Rawlings, S.: 1997, *MNRAS* **286**, 241.
Kellermann, K.I., Sramek, R., Schmidt, M., Shaffer, D.B. and Green, R.: 1989, *AJ* **98**, 1195.
Koekemoer, A.M., Henkel, C., Greenhill, L.J., et al.: 1995, *Nature* **378**, 697.
Mauersberger, R., Wilson, T.L. and Henkel, C.: 1988, *A&A* **201**, 123.
Miyoshi, M., Moran, J., Herrnstein, J., et al.: 1995, *Nature* **373**, 127.

Morganti, R., Oosterloo, T.A., Tadhunter, C.N., et al.: 2001, *MNRAS* **323**, 331.
Peck, A.B., Henkel, C., Ulvestad, J.S., et al.: 2003, *ApJ* **590**, 149.
Ponomarev, V.O., Smith, H.A. and Strelnitski, V.S.: 1994, *ApJ* **424**, 976.
Urry, C.M. and Padovani, P.: 1995, *PASP* **107**, 803.
Wilman, R.J., Edge, A.C., Johnstone, R.M., Crawford, C.S. and Fabian, A.C.: 2000, *MNRAS* **318**, 1232.

LOW-LUMINOSITY EXTRA-GALACTIC H_2O MASERS

YOSHIAKI HAGIWARA

ASTRON, Dwingeloo, The Netherlands; E-mail: hagiwara@astron.nl

(Received 16 April 2004; accepted 15 June 2004)

Abstract. Recent high-resolution observations of low-luminosity extra-galactic H_2O masers towards M 82, M 51, and NGC 4051 are presented in this contribution. These observations were conducted using the very large array (VLA). The general observational properties of these masers are discussed.

Keywords: maser-galaxies, active-galaxies, ISM-radio lines

1. Introduction

Significant advances in studies of 22 GHz extra-galactic H_2O masers have been made since the discovery of the H_2O maser towards M 33 by Churchwell et al. (1977), which was followed by the discovery of the 'nuclear' H_2O maser towards an active galactic nucleus (AGN) in NGC 4945 (Dos Santos and Lépine, 1979). In terms of isotropic luminosity, H_2O masers can be categorized into two different types: either high-luminosity ($L_{H_2O} \gtrsim 10\ L_\odot$) masers found in or near an active nucleus, or low-luminosity ($L_{H_2O} \lesssim 1\ L_\odot$) masers which are associated with star-forming regions in nearby galaxies. Milli-arcsecond VLBI imaging of nuclear H_2O masers has already proved that the high-luminosity masers associated with AGN activity are powerful tools for investigating these galaxies' innermost regions. By contrast, there has been to date no systematic study of the low-luminosity masers. It is important to study those masers at high angular resolution, as they could advance the study of eccentric extra-galactic star-formation.

2. Observations

The NRAO[1] VLA was used to observe low-luminosity H_2O masers at 22 GHz towards M 82, M 51, and NGC 4051. These observations were made between 2002 and 2003 employing a single or pair of intermediate frequency bands of width 6.25 or 12.5 MHz divided into 64 or 32 spectral channels of width 1.3 or 5.3 km s^{-1} each. The typical angular resolution of these observations was 0.1 arcsec using the VLA A-Configuration. Although the resolution is not sufficient to resolve the structure or distribution of the masers, it is crucial to measure their locations for further VLBI observations.

Astrophysics and Space Science **295:** 125–130, 2004.

References

Baudry, A., Brouillet, N. and Henkel, C.: 1994, *A&A* **287**, 20.

Baudry, A. and Brouillet, N.: 1996, *A&A* **316**, 188.

Churchwell, E., Witzel, A., Huchtmeier, W., Pauliny-Toth, I., Roland, J. and Sieber, W.: 1977, *A&A* **54**, 969.

Dietz, R.D., Smith, J., Hackwell, J.A., Gehrz, R.D. and Grasdalen, G.L.: 1986, *ApJ* **91**, 758.

Dos Santos, P.M. and Lépine, J.R.D.: 1979, *Nature* **278**, 34.

Greenhill, L.J., Moran, J.M., Reid, M.J., Gwinn, C.R., Menten, K.M., Eckart, A. and Hirabayashi, H.: 1990, *ApJ* **364**, 513.

Hagiwara, Y., Henkel, C., Menten, K.M. and Nakai, N.: 2001, *ApJ* **560**, 37L.

Hagiwara, Y., Diamond, P.J., Miyoshi, M., Rovilos, E. and Baan, W.: 2003, *MNRAS* **344L**, 53.

Henkel, C., Tarchi, A., et al.: 2004, *A&A* **414**, 117.

Ho, P.T.P., Martin, R.N., et al.: 1987, *ApJ* **320**, 663.

Kaaret, P., Prestwich, A.H., Zezas, A., Murray, S.S., Kim, D.-W., Kilgard, R.E., Schlegel, E.M. and Ward, M.J.: 2001, *MNRAS* **321L**, 29.

Matsushita, S., Kawabe, R., Matsumoto, H., Tsuru, T.G., Kohno, K., Morita, K.-I., Okumura, S.K. and Vila-Vilaro, B.: 2000, *ApJ* **545**, 107L.

Tarchi, A., Henkel, C., Peck, A.B. and Menten, K.M.: 2002a, *A&A* **385**, 1049.

Tarchi, A., Henkel, C., Peck, A.B. and Menten, K.M.: 2002b, *A&A* **389**, 39.

Weiss, A., Neininger, N., et al.: 2001, *A&A* **365**, 571.

STATISTICAL APPROACH

to not only depend on the L_{fir} of the YSO, but also on the directional properties of the jet. Because it is likely that within the Medicina beam there will be more than one YSO (as there will be within one IRAS beam), we are really investigating global properties of a cluster, rather than of a single object. In this context L_{fir} can be seen as an upper limit to the luminosity of the brightest YSO and in this sense allows discrimination between the environments of YSOs of different luminosities. Furthermore we note that while there is an established correlation between the energetics of large-scale outflows (as measured e.g. in CO) and of water masers, it is also true that the velocity extent of the bulk of the outflowing gas is typically less than 10 km s^{-1}, while maser emission can occur over a range of velocities of 100 km s^{-1} or more. This means that maser components, certainly those at large red- or blue-shifted velocities (with respect to that of the parent cloud) must be directly excited by high-velocity gas originating from the YSO. This jet-component also entrains the ambient gas, which forms the (slower-moving) large-scale outflow.

Due to space limitations only some of the results can be presented here; for a complete account, subtleties, and caveats see Brand et al. (2003).

1. From the fact that $V_{up} \approx V_{fr}$, we see that the velocity at which the maser emission is most intense is also that where it occurs most often. The representative mean maser velocity, V_{up}, is always within 7.5 km s^{-1} of the parent molecular cloud: $V_{up} - V_{cl} = -0.4 \pm 3.5$ km s^{-1}. This implies that maser emission is maximum for zero projected velocities with respect to the local environment, i.e. when the plane of the shock that creates the masing conditions is along the line-of-sight (cf. Elitzur et al., 1989).
2. $L^{up}_{H_2O}$ correlates well with the luminosity of the YSO; a power-law fit gives $L^{up}_{H_2O} = 6.4 \times 10^{-8} L_{fir}{}^{0.81 \pm 0.07}$. This is in good agreement with the slope of 1.00 ± 0.07 of a fit to the outer envelope of the distribution of single observations of many maser sources obtained by Wouterloot et al. (1995).
3. As can be seen from the *FVt*-diagrams in Figure 1, maser variability is a complex affair. Nevertheless, its overall behaviour is described by the ratio S_{max}/S_{mean} between the maximum and the mean integrated flux densities over the whole monitoring period. From Figure 2a one sees that high-luminosity YSOs tend to be associated with more stable masers. This can be understood in the context of the basic framework assumption mentioned above: around lower-luminosity YSOs a smaller number of maser components gets excited, and their intrinsic time-variability will dominate the total output; near higher-luminosity YSOs a larger number of components may be simultaneously excited, reducing the effect of their individual time-variability.
4. ΔV_{tot} is the total velocity extent of the maser emission in the frequency-of-occurrence histograms; it is shown as a function of $L^{up}_{H_2O}$ in Figure 2b. The data show that a more luminous YSOs *can* excite maser emission over a larger velocity range, but does not necessarily always do so, as this also depends on

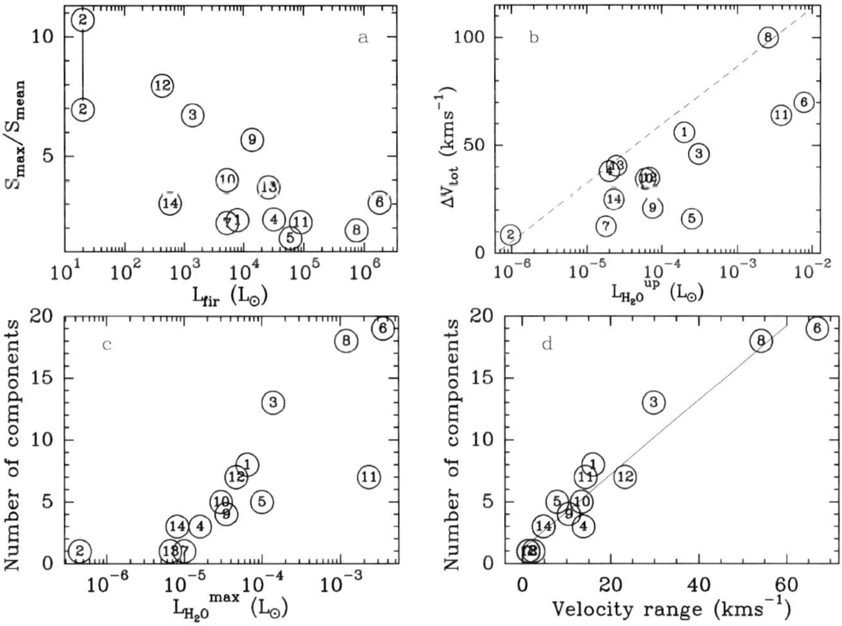

Figure 2. A selection of the results. The numbers inside the data-circles correspond to the object-numbers in Table I. (a) The ratio of maximum and mean integrated flux density as a function of the luminosity of the YSO. (b) The total velocity extent of the maser emission (from the frequency-of-occurrence histograms) as a function of the maser luminosity $L^{\mathrm{up}}_{\mathrm{H_2O}}$. (c) The number of velocity components in the spectrum with the highest integrated flux density during monitoring, as a function of $L^{\mathrm{max}}_{\mathrm{H_2O}}$, the maser luminosity determined from that same spectrum. (d) As (c), but as a function of the velocity range of the emission in the spectrum.

the local environment: the alignment of the jets with respect to the line-of-sight, and the local gas density will play a role too (see also items 9 and 10).

5. To excite all the potential sites of maser emission, sufficient energy is required to pump them. This energy is delivered by shocks, created by jets driven by the YSO. To excite maser emission at large (with respect to the parent cloud) velocity, powerful flows are needed, and hence more luminous YSOs. These will thus be associated with maser spectra containing more emission components over a larger range in velocity, as is indeed found: Figures 2c, d.
6. The FVt-diagrams can be used to study bursts of maser emission or to determine velocity drifts of individual maser components. Regarding the former, the limiting factor is our relatively large sampling interval of 60–100 days, while identifying individual velocity components and following them in time is only possible for masers with few components (cf. Mon R2 or L1204-G in Figure 1), or for components at velocities far from the crowded parts of the spectrum. From an analysis of 15 suitable maser components we found velocity gradients (both positive and negative) ranging between 0.02 and

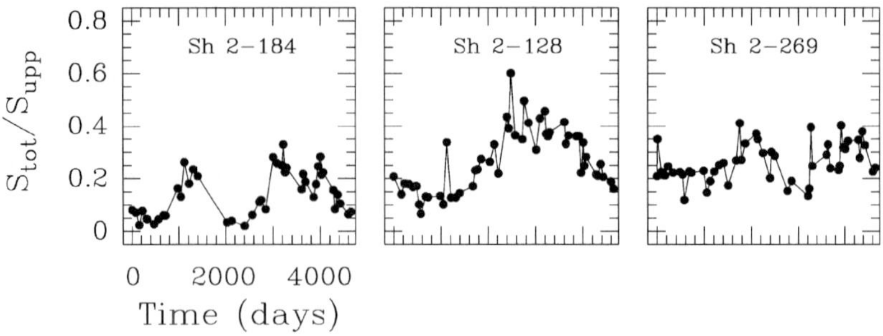

Figure 3. The integrated flux density, normalized by that of the upper envelope, as a function of time. The panels illustrate the 'super-variability' present in the maser output of the three sources shown.

1.8 km s^{-1} yr^{-1}. From an analysis of 14 bursts in 9 components in 6 sources we found durations $\Delta t \sim$ 63–900 days; the lower value is determined by our sampling interval, the higher value by the fact that bursts of longer duration are not recognized as such. Flux density increases ΔF ranged from a few tens to a few thousand percent. Both Δt and ΔF were found to decrease with increasing velocity offset of the maser component; this is consistent with what is indicated by the shape of the upper envelopes and the detection rate histograms.

7. In addition to occasional outbursts of individual maser components, in several sources a long-term, possibly periodic variability is detected in the total maser output. Figure 3 shows this for three of the best cases. The estimated duration of a full cycle varies between 5.5 (Sh 2 184, Sh 2 269) and 11 (Sh 2 128) years. This 'super-variability' might be caused by periodic variation in the YSO-power supply (which drives the jet that creates the shocks that excite the maser).
8. In Figure 4a we show the scaled velocity-time-intensity diagrams, with $L_{\rm fir}$ increasing from bottom to top, and from left to right. Note how for $L_{\rm fir} \geq 3 \times 10^4$ $L_\odot$ the emission becomes increasingly complex: from Mon R2 IRS3 to W43 Main3 the maser emission changes from being dominated by a single component to being highly structured and multi-component; the velocity extent of the emission also increases. For lower $L_{\rm fir}$ there is a variety of morphologies, but without a systematic trend with $L_{\rm fir}$, and with a smaller velocity range. The source with the lowest $L_{\rm fir}$ in the sample ($\sim$20 $L_\odot$) has not been detected most of the time.
9. Figure 4b shows the scaled frequency-of-occurrence diagrams, ordered in $L_{\rm fir}$ as in the *FVt*-diagrams. We see that for YSOs with $L_{\rm fir} \geq 3 \times 10^4$ $L_\odot$ there is at least one velocity component of the H_2O maser with a detection rate of 100%, and it is always very close to $V_{\rm cl}$. This is in agreement with what was found by Wouterloot et al. (1995) using a very different sample and observational approach. For sources with lower $L_{\rm fir}$ the typical maser detection rate (above the 5σ-level) is 75–80%. The YSO with the lowest $L_{\rm fir}$ has a detection rate of only $\sim$10%.

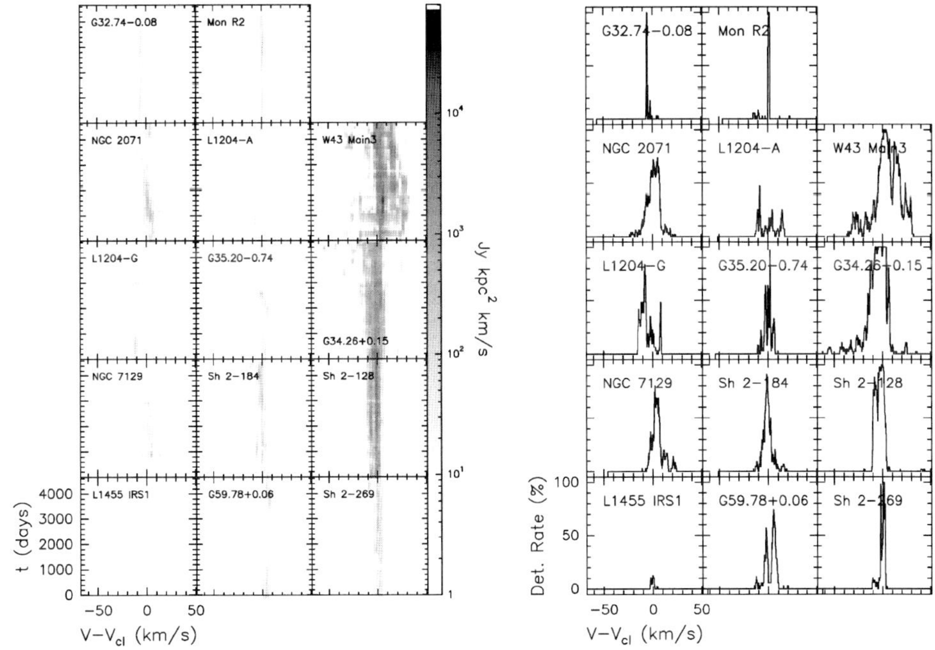

Figure 4. (a) Left: FVt-maps, like the ones shown in Figure 1, but the flux densities are scaled by d^2 and the logarithmic intensity scale is the same for all panels, thus as if the sources were at the same distance. The velocities are relative to the parent-cloud velocity $V_{\rm cl}$. The sources are ordered in $L_{\rm fir}$ of the YSO, which increases from bottom to top and from left to right. (b) Right: detection rate histograms. Ordered in $L_{\rm fir}$ as in (a).

The steep decline of the histograms with velocity away from $V_{\rm cl}$ indicates that more blue- and red-shifted maser components have shorter lifetimes than those near the systemic velocity. An interesting feature of the histograms is the 'tail', consisting of a collection of smaller peaks, which for sources with $L_{\rm fir} \geq 3 \times 10^4$ $L_\odot$ is preferentially on the blue side of the main peak (except only for Sh 2 128). As mentioned earlier, maser amplification is maximum in the plane of the shocks, where the gain path is longest (Elitzur et al., 1989). For high-velocity blue- and red-shifted components the plane of the shock is perpendicular to the line-of-sight, and the gain path is short. But if a well-collimated jet is precisely aligned with the line-of-sight, the maser can amplify the background continuum (from the HII region) and the blue-shifted components can become more intense. The strength and velocity offset of the maser feature will be determined by the jet's collimation, its alignment with the line-of-sight, and by the background radio continuum (hence by the luminosity of the YSO). Figure 4b shows that in our sample these effects combine most favourably in G34.26−0.15 and W43 Main3. High-velocity red-shifted components will always be weaker, because they cannot amplify the continuum background. If there is no outflow, or if it is driven by a lower-luminosity YSO, high-velocity components will not be seen at all. If the maser emission comes primarily from a protostellar disk, blue- and red-shifted components can be seen, but at velocities close to V_{cl}. Thus, maser emission would also be a function of the geometry of the SFR, in particular of the orientation of the beam of the outflow with respect to the line-of-sight to the YSO.

10. It appears that sources with 4×10^2 L$_\odot$ $\leq L_{fir} \leq 6 \times 10^4$ L$_\odot$ (from NGC7129 to Sh 2 269 in order of increasing luminosity) have similar (scaled) upper envelopes (not shown), with peak values of $\log[F_\nu d^2] \sim 3$ and with comparable velocity range. Outside this homogeneous group of sources we find on the one hand the lowest-luminosity source ($L_{fir} \approx 20$ L$_\odot$) with $\log[F_\nu d^2]$ two orders of magnitude smaller and with a narrower velocity range of the emission, and on the other hand the sources with higher luminosity ($L_{fir} \geq 6 \times 10^4$ L$_\odot$), with peak values of $\log[F_\nu d^2] \sim 4$–5 and a larger extent in velocity. These distinctions may reflect three different regimes of maser excitation. In the lowest luminosity sources, the maser excitation occurs on a small ($\sim$100 AU) spatial scale and might be produced by the stellar jets visible in the radio continuum. The outflows are either less powerful or impact with a lower-density ambient medium, where conditions are not suitable to create masers. In the intermediate luminosity class, the larger energetic input from the jet, as well as the presence of a higher-density molecular gas, are the main agents that determine the conditions for maser excitation. In the most luminous sources, conditions for maser excitation are similar to those in the previous category, but in this case the energetic input is so large that all potential maser sites are excited, and the determining factor is the YSO luminosity.

4. Summary

We have presented an analysis of more than 10 years-worth of water maser data of 14 SFRs. There is a clear overall dependence of the parameters of the maser emission on the YSO luminosity. In addition, we find the existence of different L_{fir} regimes and a threshold YSO-luminosity (of $\sim 10^4$ L$_\odot$) that can account for the various observed characteristics of the H_2O maser emission. Furthermore, the maser emission is found to depend on the morphology of the SFR, particularly the orientation of the outflow-jet with respect to the line-of-sight, and the density of the parent molecular environment.

For the full detailed analysis, see Brand et al. (2003).

References

Brand, J., Cesaroni, R., Caselli, P., et al.: 1994, *A&AS* **103**, 541.

Brand, J., Cesaroni, R., Comoretto, G., Felli, M., Palagi, F., Palla, F. and Valdettaro, R.: 2003, *A&A* **407**, 573.

Comoretto, G., Palagi, F., Cesaroni, R., et al.: 1990, *A&AS* **84**, 179.
Elitzur, M., Hollenbach, D.J. and McKee, C.F.: 1989, *ApJ* **346**, 983.
Valdettaro, R., Palla, F., Brand, J., et al.: 2001, *A&A* **368**, 845.
Valdettaro, R., Palla, F., Brand, J., Cesaroni, R., Comoretto, G., Felli, M. and Palagi, F.: 2002, *A&A* **383**, 266.
Wouterloot, J.G.A., Fiegle, K., Brand, J. and Winnewisser, G.: 1995, *A&A* **301**, 236; Erratum: 1997, *A&A* **319**, 360.

DENSE GAS, CHEMISTRY AND STAR FORMATION IN LUMINOUS GALAXIES

SUSANNE AALTO
Onsala Rymdobservatorium, Chalmers Tekniska Högskola, Onsala, Sweden;
E-mail: susanne@oso.chalmers.se

(Received 16 April 2004; accepted 15 June 2004)

Abstract. The molecular phase of the ISM constitutes the main source of fuel for the activity in starburst and AGNs. The physical conditions and chemical constitution of the molecular gas will change with, and respond to, the evolution of the activity. This paper includes a short discussion of the ^{12}CO/^{13}CO 1–0 line intensity ratio as a diagnostic tool of the molecular gas properties of luminous galaxies – paired with examples of high-resolution studies of how the line ratio varies within galaxies. A possible connection between the OH megamasers and galaxies with unusually high ^{12}CO/^{13}CO 1–0 line intensity ratios are also briefly discussed.

The relative intensities of the dense gas tracers HNC, HCN, HCO^+ and CN are a result of both chemistry and starburst evolution. The discussion on the interpretation of HNC 1–0 emission includes the importance of ion-neutral chemistry in a luminous starburst region. Finally, simple cartoon ISM models and how they can be applied to LIRGs and ULIRGs, are presented.

Keywords: galaxies: evolution, galaxies: ISM, galaxies: starburst, radio lines: galaxies, radio lines: ISM

1. Introduction

The inner kiloparsecs of starburst and interacting galaxies harbour vast amounts of molecular gas, 10^9–10^{10} $M_\odot$ (e.g. Scoville et al., 1991; Bryant and Scoville, 1996, 1999). In these environments, molecular clouds feed the central activity (star formation, AGNs) but are also subject to intense radiation fields, high midplane pressures, winds from supernovae, hot stars and AGNs. Therefore, understanding the conditions and chemistry of the molecular ISM is an essential part of understanding the triggering and evolution of the central activity.

Considerable efforts to go beyond observing only the lower transitions of ^{12}CO have been made in the past few years. An increasing number of single dish and synthesis observations of the isotopomer ^{13}CO reveal excitational variation among, and within, galaxies. Luminous galaxies with unusual ^{12}CO/^{13}CO line intensity ratios and warm dust indicate extreme ISM conditions in their central kiloparsec, which dominates their global CO emission. In Sections 2 and 3, typical ^{12}CO/^{13}CO 1–0 intensity ratios are discussed, as well as how to interpret the line ratio as an ISM diagnostic tool.

Astrophysics and Space Science **295:** 143–153, 2004.

In the inner kiloparsec of galaxies the mass fraction of dense ($n \gtrsim 10^4$ cm^{-3}) molecular gas is higher than in the disk of the Galaxy (e.g. Helfer and Blitz, 1993; Aalto et al., 1994). This gas phase is best traced by high dipole moment molecules, such as the polar molecule HCN (dipole moment 2.98 debye). In particular in distant luminous ($L_{\rm IR} > 10^{11}$ $L_\odot$, LIRGs) and ultraluminous ($L_{\rm IR} > 10^{12}$ $L_\odot$, ULIRGs) systems the HCN 1–0 line is the prototypical tracer of dense gas content (e.g., Solomon et al., 1992; Helfer and Blitz, 1993; Curran et al., 2000, 2001; Gao and Solomon, 2004). In Section 4, alternative high density gas tracers such as HNC, HCO^+ and CN are disussed and how their relative intensities can be interpreted in terms of the chemistry and evolution of the dense gas.

Are all starbursts intrinsically alike? Is it likely that the *compact* starburst of the ULIRG Arp 220 and the *extended* burst of the nearby, comparatively puny, starburst-galaxy M82 will evolve in a similar way? Arp 220 has a hundred times more molecular material squeezed into a smaller region than the, two orders of magnitude less luminous, starburst region of M82. The environment in which the molecular clouds live is expected to be vastly different in the two galaxies which cautions us not to treat Arp 220 as just a scaled up version of M82. In Section 5 we present two simple models for the molecular ISM of compact and extended starbursts.

2. ^{12}CO and ^{13}CO in Luminous Galaxies

To get a first handle on the conditions in the molecular ISM one can measure the $J = 1$–0 intensity ratio between the isotopomers ^{12}CO and ^{13}CO, $\mathcal{R}$. The molecular line ratio serves as a diagnostic tool for the cloud conditions of an ensemble and, given that the abundance ratio between the two does not deviate largely from that of the galaxy (30–100), the ratio provides a rough estimate of the average ^{12}CO 1–0 optical depth within the beam. The value of $\mathcal{R}$ reflects the conditions in the clouds that dominate the emission within the antenna beam, even when individual clouds are not resolved. In case two cloud components of different beam fillings and properties are present, the ratio places constraints on their relative contributions to the total emission. High values of $\mathcal{R}$ suggest low optical depth (τ) in ^{13}CO, and either low ($\tau \lesssim 1$) or moderate ($\tau \approx 1$) optical depths in ^{12}CO. Within galaxies there is a general trend of increasing $\mathcal{R}$ towards the central region where the gas is warmer and denser (e.g. Wall et al., 1993; Aalto et al., 1995; Paglione et al., 2001). Globally, there is a correlation between $\mathcal{R}$ and the FIR $f(60\ \mu\text{m})/f(100\ \mu\text{m})$ flux ratio (see Figure 3, top panel) (Young and Sanders, 1986; Aalto et al., 1991b, 1995). There are several mechanisms that may cause an elevation of $\mathcal{R}$:

1. *Moderate optical depth in the* ^{12}CO *1–0 line* caused by the gas being *warm and dense* – this will depopulate the lower levels casuing an optical depth reduction (e.g. Aalto et al., 1991a,b, 1995, 1999). The net effect on brightness temperature will be greater for the optically thin ^{13}CO than for the moderately

opaque ^{12}CO – resulting in an elevation of the intensity ratio, $\mathcal{R}$ with increasing temperature. In this scenario, we expect the ^{13}CO 2–1 line to be brighter than ^{13}CO 1–0. An example of this is the merger Arp 299 where the global ^{13}CO 1–0 emission is faint, but the ^{13}CO 2–1 emission is normal (Aalto et al., 1999; Glenn and Hunter, 2001) (see Figures 1 and 2). The ^{13}CO 2–1/1–0 line ratio exceeds 3 which is consistent with gas densities $n \gtrsim 5000$ cm^{-3} and gas kinetic temperatures $T_K \gtrsim 50$ K.

2. *Moderate optical depth in the* ^{12}CO *1–0 line* caused by the presence of *diffuse (unbound) molecular gas*. The linewidhts of the molecular clouds are large causing a reduction of the ratio of line-of-sight column density over linewidth ($N/\delta V$), resulting in a lower optical depth (e.g. Aalto et al., 1994, 1995; Downes and Solomon, 1998; Hüttemeister et al., 2000). Downes and Solomon (1998) find that many ULIRGs have such a moderate optical depth intercloud molecular medium, and Hüttemeister et al. (2000) find that a diffuse molecular medium may dominate the ^{12}CO 1–0 emission in bar-shocks in strongly barred galaxies. In this scenario we expect also the ^{13}CO 2–1 line to be relatively faint, in particular

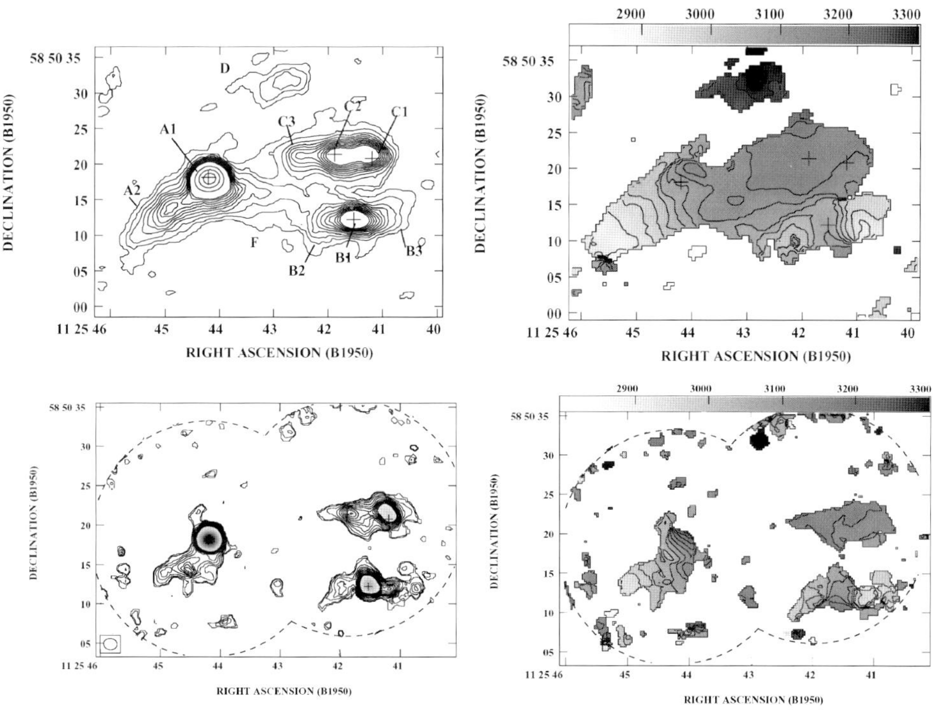

Figure 1. OVRO ^{12}CO 1–0 and 2–1 maps of the merger Arp 299. Top left: naturally weighted map of the integrated 1–0 intensity; top right: naturally weighted 1–0 velocity field; bottom left: naturally weighted map of the 2–1 integrated intensity; bottom right: the ^{12}CO 2–1 velocity field. Crosses mark radio continuum positions.

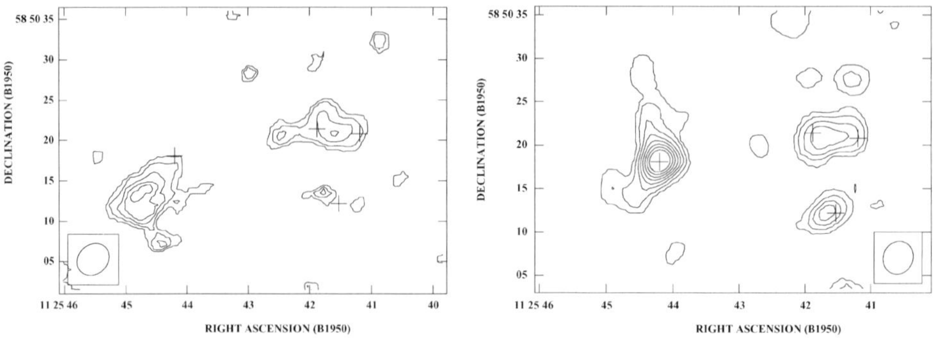

Figure 2. OVRO ^{13}CO integrated intensity maps of Arp 299. Left: ^{13}CO 1–0; right: ^{13}CO 2–1. Crosses mark radio continuum positions.

if the density of the diffuse molecular gas is $\lesssim 10^3$ cm^{-3}. This is because the ^{13}CO 2–1 line will not be thermalized at such a low density, and therefore the ^{13}CO 2–1/1–0 line intensity ratio will be less than unity.

3. *Optically thick* ^{12}CO *1–0* emission in an environment of absormally low ^{13}CO *abundances*. This has been suggested by Casoli et al., 1992 to explain the high ratio mergers. The molecular clouds are optically thick, but low metallicity gas has flown from the very outskirts of the disk to the center in the merging process. High-resolution studies so far suggest that this effect is marginal in the centers of luminous galaxies (e.g. Aalto et al., 1997, 1999) but for gas in the outskirts of galaxies, this effect may be important. Molecular gas in the outskirts of luminous major mergers will, however, not significantly affect their global ^{12}CO emission.
4. *Selective ultraviolet dissociation in PDRs* may help maintain a large ^{12}CO/^{13}CO abundance ratio. This effect is, however, not expected to be efficient if the ^{12}CO line is very opaque, but can further elevate $\mathcal{R}$ in regions where the optical depth is already moderate (e.g. Aalto et al., 1995).

In Section 5, we discuss two ISM types – the ‘raisin roll’ and ‘fried egg’ – which will both give rise to elevated values $\mathcal{R}$. The presence of diffuse and/or highly excited gas in centers of galaxies may affect the applicability of the standard ^{12}CO luminosity to H_2 mass conversion factor (the ‘X-factor’) (e.g. Aalto et al., 1994; Hüttemeister et al., 2000; Paglione et al., 2001).

2.1. Are the HR galaxies and OH megamasers the same objects?

Galaxies that are faint in the ^{13}CO 1–0 line with $\mathcal{R} \gtrsim 20$ (from now on referred to as *the high ratio (HR) galaxies*) are often interacting galaxies with FIR flux ratios $f(60\ \mu\mathrm{m})/f(100\ \mu\mathrm{m}) \gtrsim 0.78$, implying hot dust ($T_\mathrm{D} \gtrsim 30$–40 K, depending on the opacity law). The HR galaxies were first described as a sample in Aalto et al. (1991b) and their nature further discussed in Aalto et al. (1995).

Unger et al. (1986) found that galaxies with intense OH maser action have significantly different infrared colours from the rest of the sample of galaxies in which they searched for maser activity. OH maser galaxies show strong excesses at both 25 and 60 μm, which they related to the pumping process of the OH masers. They find limits to the FIR flux ratio of $f(60\ \mu\text{m})/f(100\ \mu\text{m}) \gtrsim 0.71$ for maser sources and for megamasers this is even stricter $f(60\ \mu\text{m})/f(100\ \mu\text{m}) \gtrsim 0.8$. *Thus, the* OH *megamaser galaxies occupy the same FIR colour range as the HR galaxies* (see Figure 3). As is the case for the HR galaxies, there is

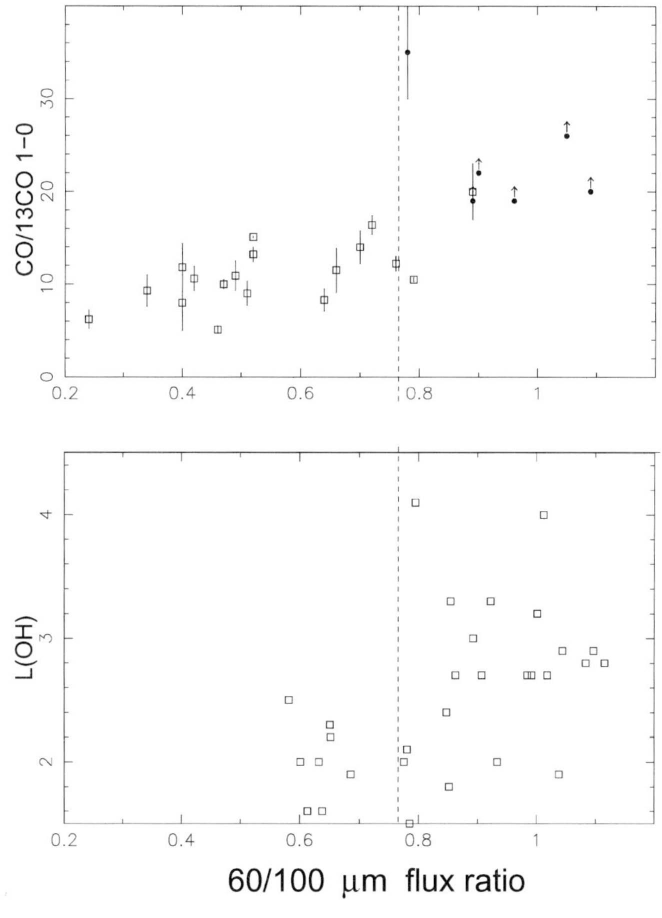

Figure 3. Top: the ^{12}CO/^{13}CO 1–0 intensity ratio, towards a sample of gas-rich luminous galaxies, plotted versus the far-IR flux ratio at 60 μm and 100 μm. Error bars are 1σ and lower limts 2σ. Filled symbols indicate galaxies with a total far-IR luminosity $L_{\text{FIR}} \gtrsim 3 \times 10^{11}\ L_{\odot}$. Bottom: the OH megmaser luminosity plotted against the FIR $f(60\ \mu\text{m})/f(100\ \mu\text{m})$ flux ratio. The dashed line marks the $f(60\ \mu\text{m})/f(100\ \mu\text{m})$ flux ratio of 0.78.

an overrepresentation of dynamically disturbed objects among OH megamaser sources.

Randell et al. (1995) present molecular cloud models that will reproduce the 1667/1665 MHz OH maser emission: small clouds (3 pc); medium to dense gas "a few times 10^3 to a few times 10^4 cm^{-3}"; high dust and gas temperatures: $T_d > T_k > 40$–50 K. These conditions are suggested to prevail in OH-masing clouds in the inner 50–200 pc ("the OH zone") of OH megamaser galaxies. *The resulting* ^{12}CO/^{13}CO *1–0 intensity ratio of such an ensemble will be high for a large range of* ^{12}CO/^{13}CO *abundance ratios*. This is because both lines are thermalised at those densities and the high kinetic temperature will then result in a depopulation of the lower levels, reducing the optical depth. Interestingly, the cloud models for the inner kiloparsec of IC 694 (the eastern galaxy in the merger Arp 299), based on the combined OVRO high-resolution data on ^{12}CO 1–0, 2–1, ^{13}CO 1–0, 2–1 and HCN 1–0, yield a very similar result as the models presented by Randell (Aalto et al., 1999). *We therefore suggest that the OH megamaser galaxies are a part/subset of the HR galaxies.* The OH megamaser galaxies have a significant fraction of their molecular gas mass in the inner few hundred pc where the intense IR field from the warm dust pumps the OH maser action. A further discussion will appear in Aalto and Polatidis (in preparation). In the Randell models, larger clouds (25 up to 300 pc) may also show 18 cm emission, but weaker. In galaxies where some of the molecular emission emerges from diffuse, unbound clouds also this phase of the ISM may partake in the 1667/1665 MHz masing – or may show up in absorption. As we shall see later, the ULIRGs may have a significant fraction of its lower transition ^{12}CO emission emerging from diffuse unbound molecular clouds.

2.2. THE NON-MASING HR GALAXIES

The HR galaxies can roughly be divided into two subgroups: the HCN-bright with very compact molecular distributions – and the HCN-faint with less compact molecular distribution. The former sample is, on average, a factor of a few times more luminous than the latter. Arp 220, Mrk 231, IC 694, NGC 2623, NGC 6240 all belong to the former class with ^{12}CO/HCN 1–0 intensity ratios between 3 and 10. They have compact ^{12}CO distributions with a very significant fraction of the emission concentrated in the inner 400 pc – and a high mass fraction of dense gas. The second subgroup (NGC 4194, NGC 1614 and NGC 3256) have in general ^{12}CO/HCN 1–0 intensity ratios >20. They are all starbursts with more extended molecular and burst regions of one to a few kpc, and a relatively low mass fraction of dense gas. In this class of galaxies, the molecular cloud distribution is likely not compact enough for efficient pumping of the OH masers. (For an ISM model of NGC 3256 see Aalto et al., 1991a.)

3. Dense Gas in Galaxies – Star Formation and Chemistry

3.1. HCN-HNC IN GALAXIES

Solomon et al. (1992) and Gao and Solomon (2004) find a tighter correlation between FIR and HCN luminosity than the one found between FIR and CO. They suggest that, in general, this implies that the IR luminosities originate from star formation rather than AGN activity in FIR luminous galaxies. In order to understand the activities of the dense gas in the centers of luminous galaxies it is however essential to go beyond correlation diagrams and attempt to measure the prevailing conditions of the dense gas. Apart from observing higher transitions of HCN this can be achieved through observing the emission from other high density tracers. One such tracer is the HNC molecule, the isomer of HCN. In a chemistry dominated by neutral–neutral reactions, it is predicted that the $\frac{\mathrm{HCN}}{\mathrm{HNC}}$ ratio increases with increasing temperature and gas density (e.g., Schilke et al., 1992) and that a bright HNC 1–0 line is an indicator of the presence of significant masses of cold (10 K) dense gas.

Aalto et al. (2002) undertook a survey of HNC 1–0 line emission in a sample of 13 luminous IR galaxies ($L_{\mathrm{IR}} > 10^{11}\ L_{\odot}$) with previously measured HCN luminosities. They found that the $\frac{\mathrm{HCN}}{\mathrm{HNC}}$ 1–0 ratios vary strongly within the sample – from 1 to $\gtrsim$6. From this we can learn that the actual *properties* of the dense gas vary significantly from galaxy to galaxy, even if their FIR luminosities, HCN luminosities (and global ^{12}CO/HCN 1–0 luminosity ratios) are similar.

Aalto et al. (2002) furthermore find that the IR luminous galaxies are at least as HNC luminous as more nearby (IR-fainter) galaxies (Hüttemeister et al., 1995). Galaxies, like Mrk 231, where there are strong still showed bright HNC 1–0 lines. Aalto et al. (2002) conclude that the *HNC emission is not a reliable tracer of cold (10 K) gas in the center of luminous IR galaxies*, the way it may be in clouds in the disk of the Milky Way. Instead, several alternative explanations were considered, including optical depth effects and a chemistry dominated by ion-neutral reactions (instead of neutral–neutral reactions). We learn that standard interpretations based on molecular clouds in the disk of the Milky Way cannot be used for the inner kiloparsec of luminous galaxies. When the gas densities are not too high ($n = 10^4$–10^5 cm^{-3}) and the ion abundance is large enough, then reactions with $\mathrm{HCNH^+}$ (HCN and HNC reacts with $\mathrm{H_3^+}$ to form $\mathrm{HCNH^+}$) become more important. Once HCN and HNC become protonised, $\mathrm{HCNH^+}$ will recombine to produce either HCN or HNC with 50% probability. At higher densities, the ion abundance is likely lower and reactions like $\mathrm{HNC + O \rightarrow CO + NH}$ become more important at high temperatures. Therefore, when interpreting the HCN/HNC line ratio the chemical aspect of the starburst evolution must also be considered. It is easy to conceive of a chemistry dominated by fast ion-neutral reactions at the heart of an active starburst where there is strong interaction between the activity itself (in photon dominated regions, PDRs) and the surrounding ISM.

For a further discussion of ion-neutral and neutral–neutral chemistry, please see papers by Spaans or Hogerheijde in this volume.

3.1.1. *How to Interpret a Large HCN/HNC Ratio in the Ion-Neutral Chemistry?* A few of the very luminous galaxies however, show large HCN/HNC 1–0 line ratios (e.g. Mrk 273). Is that a result that can be interpreted in the context of ion-neutral chemistry? Either the gas is very dense and warm and reactions are governed by neutral–neutral chemistry. The starburst is thus in an early phase dominated by undisrupted, very dense and warm gas – a pre-PDR phase. Alternatively, HNC is removed by shocks in a later stage of the starburst. To distinguish between these two scenarios we need more information, such as observations of shock tracer molecules and higher transition observations of HNC/HCN.

3.2. CN AND PDRs IN LUMINOUS GALAXIES

The radical CN is another tracer of dense gas, with a somewhat lower (by a factor of 5) critical density than HCN. Observations of the CN emission towards the Orion A molecular complex (Rodriguez-Franco et al., 1998) show that the morphology of the CN emission is dominated by the ionization fronts of the H II regions. The authors conclude that this molecule is an excellent tracer of regions affected by UV radiation. Thus, the emission from the CN molecule should serve as a measure of the relative importance of gas in PDRs).

Aalto et al. (2002) also conducted a CN 1–0 (and 2–1) survey in the same galaxies they surveyed for HNC (see above). The goal was to probe the PDR phase of the molecular ISM of which CN should be a reliable tracer. Also in this study, the $\frac{\mathrm{HCN}}{\mathrm{CN}}$ 1–0 intensity ratios show significant variation – ranging from 0.5 to $\gtrsim$6. There is a surprising trend of *decreasing* CN luminosity with increasing FIR luminosity (but this trend needs to be confirmed with a larger sample). This is unexpected, since the luminosity of ULIRGs is suspected to be largely dominated by massive starbursts. It is interesting that also another classical PDR tracer is weak in ULIRGs: the [C II] 158 μm fine structure line is found to be abnormally faint compared to other, less FIR luminous, starburst galaxies like NGC 3690 (e.g., Luhman et al., 1998). Malhotra et al. (1997) report a decreasing trend in $\frac{F_{[\mathrm{CII}]}}{F_{\mathrm{FIR}}}$ with increasing $\frac{f(60)}{f(100)}$ μm flux ratio. Several possible explanations for the [C II] faintness are brought forward (e.g., Malhotra et al., 1997; Luhman et al., 1998; van der Werf, 2001). The PDRs may be quenched in the high pressure, high density environment in the deep potentials of the ULIRGs and the HII regions exist in forms of small-volume, ultracompact HII regions that are dust-bounded. The [C II] line may become saturated either in low density ($n \propto 10^2$ cm^{-3}) regions of very high UV fields ($G_0 \propto 10^3$) or in dense ($n \propto 10^5$ cm^{-3}) regions of more moderate UV fields ($G_0 = 5$–10). A soft UV field from an aging starburst is another possibility. A higher dust-to-gas ratio would also decrease the expected $\frac{F_{[\mathrm{CII}]}}{F_{\mathrm{FIR}}}$ ratio.

There is not a one-to-one correspondence between CN and [C II] faint galaxies, but the general trend is the same. This further emphasizes the need for care when using nearby, significantly less luminous, systems as prototypes for all starbursts. The ISM of the ULIRGs is doing something different than that of more modest bursts and we still have to find out whether the cause is evolutionary or whether the bursts are intrinsically different in the ULIRGs (see Aalto et al., 2002). The presence of an AGN may also significantly affect the properties of the nuclear ISM. Recent work by e.g. Kohno et al. (2001) reveal that there are also differences between AGN-driven molecular ISMs and those that are mostly associated with star forming activities. They have found that a large HCN/HCO^+ 1–0 line intensity ratio typifies an X-ray, AGN-driven chemistry (see also Lepp and Dalgarno, 1996).

4. Dense and Diffuse Gas in Compact and Extended Starbursts

In Table I we list typical line ratios for luminous, compact (high pressure) (ULIRGs) and extended lower luminosity (low pressure) starbursts. Both types may have large values of $\mathcal{R}$, but their average ISM properties differ on several accounts. In Figure 4 we show cartoons of the two ISM types, "raisin roll" and "fried eggs" (see also Aalto et al., 2001). For the former (ULIRG) scenario, the ^{12}CO emission is emerging from low density $n \lesssim 10^3$ cm^{-3} diffuse gas of large linewidths and filling factor, while the HCN 1–0 emission is coming from dense ($n \gtrsim 10^4$ cm^{-3}), embedded clouds (but

TABLE I

Line ratios in HR galaxies

Line	High pressure (compact)	Low pressure (extended)
CO 2–1/1–0	Subthermal ($\lesssim$0.7)	Thermalized ($\approx$1)
$\mathcal{R}_{1-0}$	$\gtrsim$20	$\gtrsim$20
CO/HCN	8	$\gtrsim$25

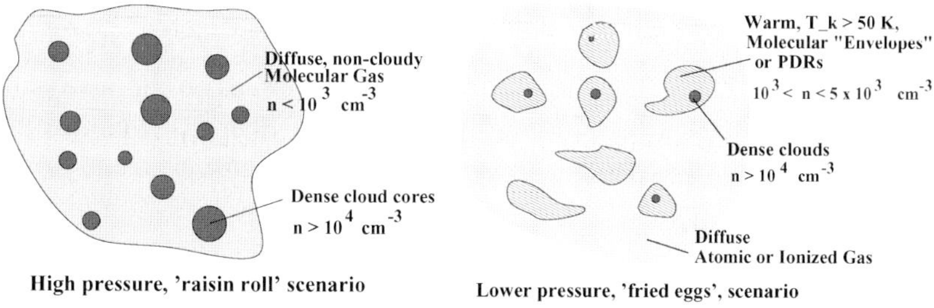

Figure 4. Models of a high pressure and low pressure ISM. In the raisin roll scenario the diffuse gas is molecular, but with a fairly low mass fraction. In the lower pressure scenario, the radiation is successfully competing with the pressure and the diffuse gas will be atomic or ionized.

see e.g. Aalto et al., 1995 for a discussion on mid-IR pumping of HCN). The bulk of the molecular mass is in the dense gas, but the high value of $\mathcal{R}$ can primarily be attributed to the large linewidths of the diffuse molecular gas (which can remain molecular because of the extreme pressure). If the dense clumps are warm, it is also likely that $\mathcal{R}$ is high also in this component of the gas. Galaxies which may be partly characterised by this ISM scenario include Arp 220, NGC 6240 and VII Zw 31. Higher transition ^{12}CO will emerge primarily from the dense clumps – and may be quite intense if the dense gas is warm. Diffuse molecular gas at the center of ULIRGs have also been discussed by Downes and Solomon (1998).

It may seem surprising that the ^{12}CO emitting gas is somewhat denser, $10^3 \lesssim n \lesssim 5 \times 10^3$ cm^{-3}, in the lower pressure scenario (and with a smaller volume filling factor). This may be because (in contrast to the "raisin roll" ISM) the intercloud medium is likely atomic (or ionized) since the destructive forces of the newborn stars are not balanced by a high ambient pressure. Thus, the ^{12}CO emitting gas is more "cloudy" in the lower pressure model, but the relatively faint HCN emission indicates that the mass fraction of *high* density gas, $n > 10^4$ cm^{-3}, is lower. The lower transition ^{12}CO emission may, however, be filamentary and not always in bound clouds. The large value of $\mathcal{R}$ is caused by elevated gas temperatures ($T_{\mathrm{k}} \gtrsim 50$ K) since the gas is dense enough for ^{12}CO to be thermalized ($T_{\mathrm{ex}} \approx T_{\mathrm{k}}$). The clouds may be PDRs if the impact of the starburst is strong enough. Galaxies which may be partly characterised by this ISM scenario include the Medusa (NGC 4194), UGC 2866 and NGC 1614.

5. Concluding Remarks

Simple models such as the ones above must of course be used with care – and should be refined and modified for each indvidual case. For example, the ISM of the nucleus of IC 694 appears to be a modified raisin roll model where the diffuse molecular gas may have been eaten away by an intense radiation field. The models should therefore be viewed as a first attempt at a general ISM characterization of starbursts of various types – and which should help us find new and better tracers of starburst chemistry and evolution.

The observations of bright emission from HNC, CN, HCO^+ and HC_3N in luminous galaxies shows that we are already able to address issues on chemistry and evolution beyond the most nearby galaxies. It is, however, clear that we must improve extragalactic ISM and PDR models to fit existing, and future, observational data.

References

Aalto, S., Black, J.H., Booth, R.S. and Johansson, L.E.B.: 1991a, *A&A* **247**, 291.
Aalto, S., Johansson, L.E.B., Booth, R.S. and Black, J.H.: 1991b, *A&A* **249**, 323.
Aalto, S., Booth, R.S., Black, J.H. and Johansson, L.E.B.: 1994, *A&A* **286**, 365.

Aalto, S., Booth, R.S., Black, J.H. and Johansson, L.E.B.: 1995, *A&A* **300**, 369.
Aalto, S., Radford, S.J.E., Scoville, N.Z. and Sargent, A.I.: 1997, *ApJ* **475**, L107.
Aalto, S., Radford, S.J.E., Scoville, N.Z. and Sargent, A.I.: 1999, in: J.E. Barnes and D.B. Sanders (eds.), *IAU Symposium 186 on Galaxy Interactions at Low and High Redshift*, p. 231.
Aalto, S., Hüttemeister, S. and Walter, F.: 2001, in: J.H. Knapen, J.E. Beckman, I. Shlosman and T.J. Mahoney (eds.), *Proceedings of the ASP Conference on the Central Kiloparsec of Starbursts and AGN: The La Palma Connection*, Vol. 249, p. 626.
Aalto, S., Polatidis, A.G., Hüttemeister, S. and Curran, S.J.: 2002, *A&A* **381**, 783.
Bryant, P.M. and Scoville, N.Z.: 1999, *AJ* **117**, 2632.
Bryant, P. and Scoville, N.Z.: 1996, *ApJ* **457**, 678.
Casoli, F., Dupraz, C. and Combes, F.: 1992, *A&A* **264**, 55.
Casoli, F., Willaime, M.-C., Viallefond, F. and Gerin, M.: 1999, *A&A* **346**, 663.
Curran, S.J., Aalto, S. and Booth, R.S.: 2000, *A&AS* **141**, 193.
Curran, S.J., Polatidis, A.G., Aalto, S. and Booth, R.S.: 2001, *A&A* **373**, 459.
Downes, D. and Solomon, P.M.: 1998, *ApJ* **507**, 615.
Gao, Y. and Solomon, P.M.: 2004, *ApJ* 606, 271.
Glenn, J. and Hunter, T.R.: 2001, *ApJS* **135**, 177.
Helfer, T.T. and Blitz, L., 1993, *ApJ* **419**, 86.
Hüttemeister, S., Henkel, C., Mauersberger, R. et al.: 1995, *A&A* **295**, 571.
Hüttemeister, S., Aalto, S., Das, M. and Wall, W.F.: 2000, *A&A* **363**, 93.
Kohno, K., Matsushita, S., Vila-Vilaro, B., Okumura, S.K., Shibatsuka, T., Okiura, M., Ishizuki, S. and Kawabe, R.: 2001, in: J.H. Knapen, J.E. Beckman, I. Shlosman and T.J. Mahoney (eds.), *Proceedings of the ASP Conference Series on Central Kiloparsec of Starbursts and AGN: The La Palma Connection*, Vol. 249, p. 672.
Lepp, S. and Dalgarno, A.: 1996, *A&A* **306**, L21.
Luhman, M.L., Satyapal, S., Fischer, J. et al.: 1998, *ApJ* **504**, L11.
Malhotra, S., Helou, G., Stacey, G. et al.: 1997, *ApJL* **491**, 27.
Meier, D.S., Turner, J.L. and Hurt, R.L.: 2000, *ApJ* **531**, 200.
Paglione, T.A.D., Wall, W.F., Young, J.S., Heyer, M.H., Richard, M., Goldstein, M., Kaufman, Z., Nantais, J. and Perry, G.: 2001, *ApJ* **135**, 183.
Randell, J., Field, D., Jones, K.N., Yates, J.A. and Gray, M.D.: 1995, *A&A* **300**, 659.
Rodriguez-Franco, A., Martin-Pintado, J. and Fuente, A.: 1998, *A&A* 329, 1097.
Schilke, P., Walmsley, C.M., Pineau de Forêts, G. et al.: 1992, *A&A* **256**, 595.
Scoville, N.Z., Sargent, A.I., Sanders, D.B. and Soifer, B.T.: 1991, *ApJ* **366**, L5.
Solomon, P.M., Downes, D. and Radford, S.J.E.: 1992, *ApJ* **387**, L55.
Unger, S.W., Chapman, J.M., Cohen, R.J., Hawarden, T.G. and Mountain, C.M.: 1986, *MNRAS* **220**, 1.
van der Werf, P.P.: 2001, in: D. Lutz and L.J. Tacconi (eds.), *Proceedings of the Ringberg Workshop on Starburst Galaxies – Near and Far*.
Wall, W.F., Jaffe, D.T., Bash, F.N. et al.: 1993, *ApJ* **414**, 98.
Young, J.S. and Sanders, D.B.: 1986, *ApJ* **302**, 680.

A MOLECULAR RING IN THE LINER NGC 5218

E. OLSSON[1], S. AALTO[2] and S. HÜTTEMEISTER[3]
[1]*Onsala Space Observatory & Jodrell Bank Observatory, Onsala, Sweden;*
E-mail: evert@oso.chalmers.se
[2]*Onsala Space Observatory, Onsala, Sweden*
[3]*Astronomisches Institut der Universität Bochum, Germany*

(Received 16 April 2004; accepted 15 June 2004)

Abstract. High resolution OVRO CO 1–0 observations of the inner 30″ in the LINER galaxy NGC 5218 reveal the presence of a double centrally peaked molecular concentration with extensions out to a radius of 12″. The molecular mass detected is 2.4×10^9 $M_\odot$ and the gas surface density is high, $\sim$3000 $M_\odot$ pc square in the inner 500 pc. The SFR is 2–3 $M_\odot$ yr^{-1} and the SFE is $\sim$13, which are low or moderate values for that gas surface density. We interpret the inner feature as a rotating molecular ring with a radius of $\sim$ 200 pc. We furthermore suggest that the LINER activity in NGC 5218 is not caused by an aging starburst, but by a buried AGN.

1. Introduction

It has long been known that certain galaxies have nuclear emission-line spectra that are not typical for either Starburst or Seyfert galaxies, but have similarities with both. If certain criteria of the intensity ratios of the emission-lines are met, the galaxy is defined as containing a low-ionisation nuclear emission-line region (LINER). There is still no single answer to what is powering the emission. To the contrary, the power sources are believed to be different for different LINERs. We have used OVRO to study the molecular gas in NGC 5218, which is an SBb pec galaxy hosting a LINER at its center. The galaxy is barred, has a high FIR luminosity ($L_{\mathrm{IR}} = 3.1 \times 10^{10}$ $L_\odot$) and is in distant interaction with the early type galaxy NGC 5216. The systemic velocity is 2880 km s^{-1}. The goal was to study how much gas that can still be found in thc large-scale bar and arms, as well as the morphology and kinematics of the gas in the inner region.

2. Observations

The Owens Valley Radio Observatory (OVRO) has been used in its low and high resolution configuration to map NGC 5218 in CO 1–0 in the inner 30″ ($R = 5.7$ kpc). The low and high-resolution synthesized beams were $\sim$4″ and $\sim$2″, respectively. The primary beam was 60″. Typical system temperatures were 400 K. The rms noise in the low- and high-resolution maps is 18 and 10 mJy per beam, respectively. The Onsala 20 m telescope was used in April and May 2003 to obtain HCN and HCO^+ 1–0 data toward the OVRO CO peak (α $13^h30^m27.2^s$, δ 63°01′27.5″, B1950) of NGC 5218. Typical system temperatures were 350 K.

3. Results and Discussion

3.1. The Low-Resolution Maps

We detect a bar-like central CO-feature (see Figure 1) with a size of 18″ × 10″ (3.5 × 2.0 kpc). The position angle is 78°. Note that the CO distribution is asymmetric with respect to the central peak. The total integrated flux is 133 Jy km s^{-1}. A

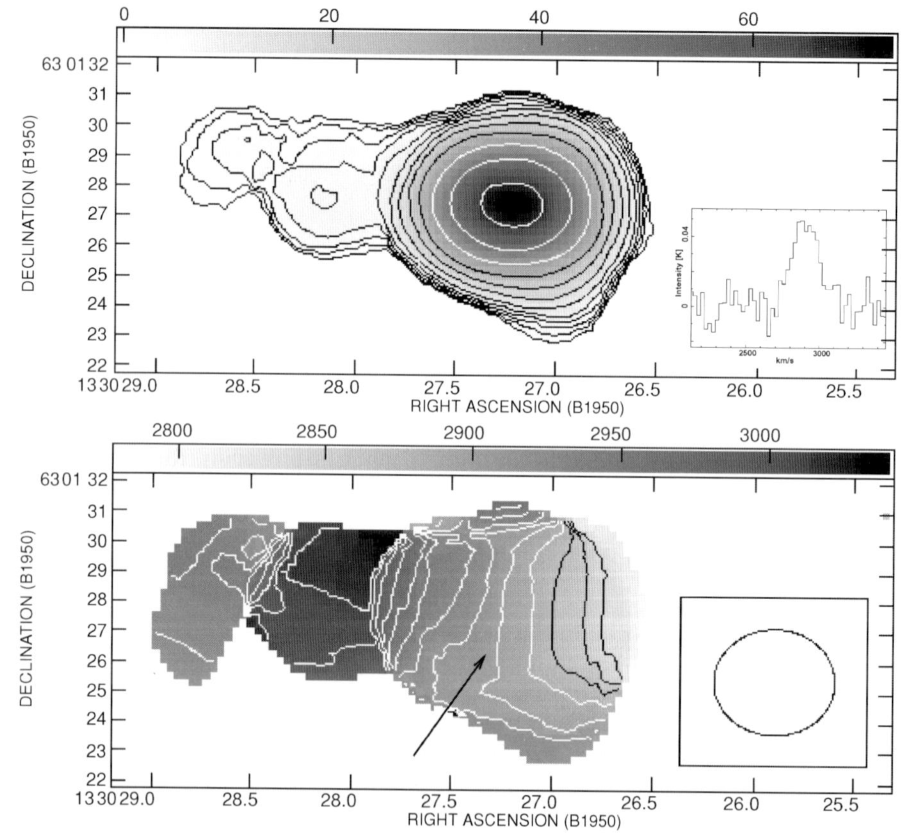

Figure 1. OVRO 4″ resolution CO 1–0 integrated intensity (upper) and velocity field (lower). The integrated intensity contour levels are 2, 2.8, 4, 5.5, 8, 11.3, 16, 22.6, 32, 45.3, 64 and 90.5% of the peak value of 77 Jy km s^{-1}. The CO spectrum is taken with the OSO 20 m telescope toward the OVRO CO peak. The arrow points at the velocity contour of 2890 km s^{-1}, and velocity contour increments are 15 km s^{-1}.

standard CO to H_2 conversion factor was used to calculate a total molecular mass of 2.4×10^9 $M_\odot$. The maximum flux in one beam (3.″75 × 3.″54) is 77 Jy km s^{-1}. This is 58% of the total flux, and yields a deprojected (for $i \sim 45°$) surface density of ∼3000 $M_\odot$ pc^{-2} in the central beam.

The velocity field map indicates that the major part of the CO-structure is in solid body rotation, except the eastern most part. Velocities range from 2800 to 3100 km s^{-1}, with the western part approaching and the eastern part receding. The part that is in solid body rotation is roughly circular (from our viewing angle), with a radius of ∼1 kpc. Velocity gradient in solid body is ∼100 km s^{-1} per kpc.

A pv-cut along the major axis (PA = 78°) also reveals solid body rotation out to approximately 1 kpc from center (Figure 3). The velocity reaches a peak at a galactocentric radius of about 1 kpc and turns over to differential rotation, or even falls.

3.2. The high-resolution maps

The high-resolution OVRO CO 1–0 maps are shown in Figure 2. We detect a symmetric double CO peak, with ∼2″ separation. The velocity field indicates solid body rotation, but with an additional kinematical component at the eastern peak. The projected rotational velocities are ∼50 km s^{-1} at a radius of ∼1″ (which implies an enclosed dynamical mass of $\sim 1 \times 10^8$ $M_\odot$).

3.3. The dense gas tracers, HCN and HCO^+

In Figure 3, we have marked the line width of the central, single-dish HCN and CO (also obtained with the OSO 20 m telescope) emission. Since the center velocity of the HCN emission is close to the systemic velocity, we speculate it is emerging from the inner 1–2 arcseconds (200–400 pc). The HCO^+ emission is fainter than HCN in terms of peak intensity by a factor of two, but may be broader with a fitted line width similar to that of CO.

3.4. Molecular mass, SFR and SFE

A standard CO-flux to H_2-mass conversion factor was used to estimate a total molecular mass of 2.4×10^9 $M_\odot$. This number must be taken as an estimate, mainly since the SCF might overestimate the gas mass, especially in centers of galaxies (Dahmen et al., 1998).

The star formation rate (SFR) as calculated from $\mathrm{SFR} = 7.7 \times 10^{-11} L_{\mathrm{FIR}}$ indicates an SFR of 2–3 $M_\odot$ yr^{-1} in NGC 5218. The ratio of the abundance of massive young stars, to the available amount of star forming material, M_{H_2} gives the star formation efficiency. In NGC 5218 this ratio is ∼13, which is a moderate value for an interacting galaxy (e.g. Young et al., 1989; Planesas et al., 1997).

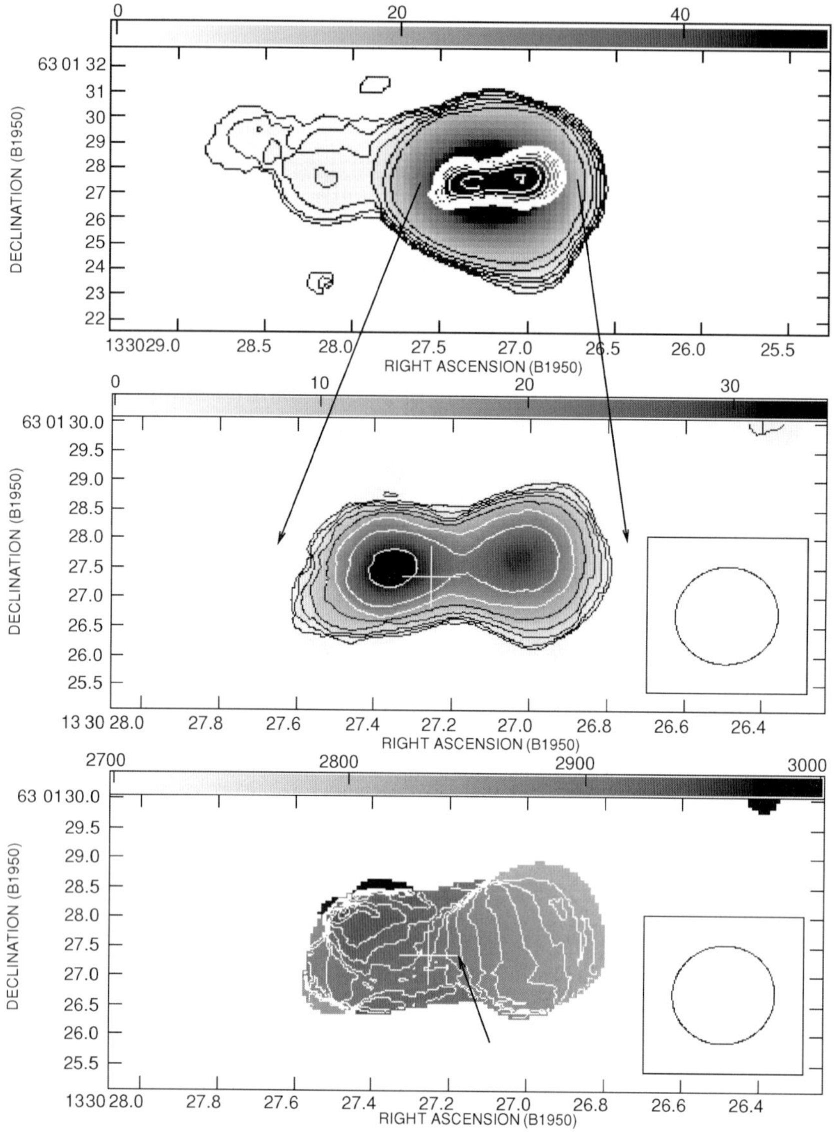

Figure 2. OVRO 2″ resolution CO 1–0 maps. Upper panel shows high-resolution contours (white) overlayed on the low (4″) resolution map. The high resolution integrated intensity contour levels are 8, 11.3, 16, 22.6, 32, 45.3, 64 and 90.5% of the peak value of 34 Jy km s^{-1}. The arrow points at the velocity contour of 2890 km s^{-1}, and velocity contour increments are 10 km s^{-1}.

3.5. The double CO peak

Bars are instrumental in letting gas flow to galactic centers. It is most likely that more than half of the molecular gas within the central kiloparsec of a barred galaxy was transported there from outside by the bar (Sakamoto et al., 1999).

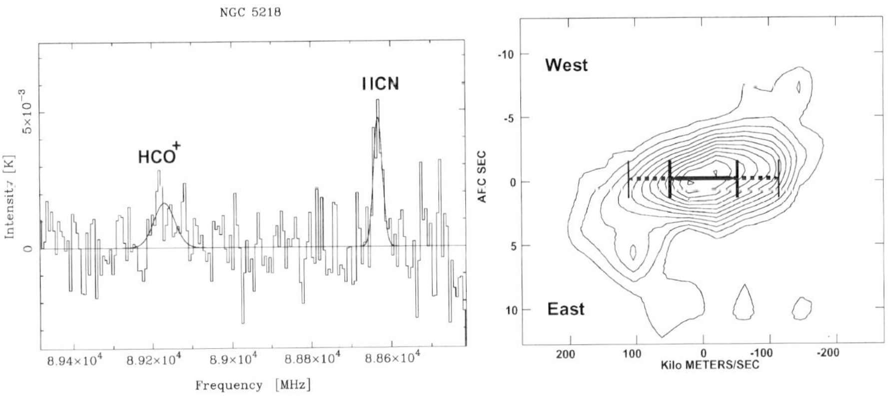

Figure 3. The OSO 20 m HCN/HCO^+ single dish spectrum and the CO pv diagram of NGC 5218 with the central single dish line widths (full width half maximum, obtained by Gaussian fits) of CO (dashed line) and HCN (filled line) indicated. The contours are at 1, 2, 3, . . ., 12 times 2 σ, where σ is 18 mJy per beam. The reference point is at the CO integrated intensity peak.

We suggest that the gas transported inwards by the $18'' \times 10''$ bar is piled up at a radius of $\sim 1''$, where we see the two peaks in the OVRO high-resolution map (Figure 2), which may be the observational response to a ring of dense molecular gas.

3.6. THE DENSE GAS—A ROTATING MOLECULAR RING?

In Figure 3 we indicate the velocity width of the HCN emission. We conclude that it must be emerging from the very inner region—the central 200 pc. Since the HCN 1–0 line requires densities in excess of $n > 10^4$ cm^{-3} to become excited, it seems like the dense gas is even more concentrated toward the center than CO—even though the gas surface density indicated by the CO emission is already quite large. The dense gas appears to be associated with what we suggest above is a rotating molecular ring. *We also suggest that the nuclear dense gas is associated with AGN activity and not star formation.* This is because the lack of HCO^+ 1–0 emission from the same region as the HCN. According to Lepp and Dalgarno (1996) a deficiency of HCO^+ is expected near a hard X-ray source, while the HCN abundance may instead increase. Hence, in an AGN dominated nuclear ISM, one expects a brighter HCN line with respect to HCO^+ than in a softer starburst environment. Together with the apparent lack of star formation activity, the structure of the HCN emission, and the weak HCO^+ emission support the notion that the LINER activity of NGC 5218 is driven by an AGN, rather than a starburst.

4. Conclusions

We detect a continuous molecular gas bar with a size of $18'' \times 10''$ (3.5×2.0 kpc). The total molecular gas mass, based on a standard galactic conversion factor, is

2001) shows broad lines (300–400 km s^{-1}) against two compact radio components. Subsequent VLBA observations of these components (Gallimore and Beswick, 2004) show them to be AGN, which has been confirmed by X-ray measurements (Komosa et al., 2003).

Clearly each object requires careful study to determine the roles of AGN and starbursts in their energetic. The problem is made more difficult by the existence of compact, highly luminous radio supernovae such as $41.95 + 576$ in M82, and the objects seen in Arp220. Although the 20 cm flux density of $41.95 + 576$ is now less than 100 mJy, it was almost 1 Jy shortly after its discovery in the 1960s, and hence would currently be detectable at distances >100 Mpc. Hence if a galaxy similar properties to M82 was situated at ~30 Mpc, then any compact variable radio source would almost certainly be given AGN status. Only by careful study of the galaxy dynamics (i.e. showing the compact source not to be at the dynamical centre) or by monitoring its monotonic flux decay over several decades, could its identity as a radio supernova be established. It therefore seems likely that a number of compact radio supernovae will have already been misidentified as AGN in the literature and hence classifications of AGN and starbursts should be treated with some caution.

Acknowledgements

I wish to thank Rob Beswick, Tom Muxlow, Phil Diamond, Karen Wills, Evert Olsson, Carole Mundell and Anita Richards for assistance with this paper.

References

Antonucci, R.R.J. and Miller, J.S.: 1985, *ApJ* **297**, 621.
Bahcall, J.N. and Ekers, R.D.: 1969, *ApJ* **157**, 1055.
Beswick, R.J., Pedlar, A., Mundell, C.G. and Gallimore, J.F.: 2001, *MNRAS* **325**, 151.
Beswick, R.J., Pedlar, A. and McDonald, A.R.: 2002, *MNRAS* **335**, 1091.
Carilli, C.L., Wrobel, J.M. and Ulvestad, J.S.: 1998, *AJ* **115**, 928.
Carilli, C.L. and Taylor, G.B.: 2000, *ApJ* **532**, L95.
Cole, G.H.J., Pedlar, A., Mundell, C.G., Gallimore, J.F. and Holloway, A.J.: 1998, *MNRAS* **301**, 782.
Gallimore, J.F., Baum, S.A., O'Dea, C.P., Brinks, E. and Pedlar, A.: 1996, *ApJ* **462**, 240.
Gallimore, J.F., Baum, S.A., O'Dea, C.P., Pedlar, A. and Brinks, E.: 1999, *ApJ* **524**, 684.
Gallimore, J.F. and Beswick, R.: 2004, *ApJ* **127**, 239.
Knapen, J.H., Laine, S., Yates, J.A., Robinson, A., Richards, A.M.S., Doyon, R. and Nadeau, D.: 1997, *ApJ* **490**, L29.
Komossa, S., Burwitz, V., Hasinger, G., Predehl, P., Kaastra, J.S. and Ikebe, Y.: 2003, *ApJ* **582**, L15.
Klöckner, H.R. and Baan, W.A.: 2002, in: *Proceedings of the Sixth European VLBI Symposium*, p. 175.
Klöckner, H.R., Baan, W.A. and Garrett, M.A.: 2003, *Nature* **421**, 821.
Kukula, M.J., Pedlar, A., Baum, S.A. and O'Dea, C.P.: 1995, *MNRAS* **276**, 1262.

McDonald, A.R., Muxlow, T.W.B., Wills, K.A., Pedlar, A. and Beswick, R.J.: 2002, *MNRAS* **334**, 912.

McDonald, A.R., Muxlow, T.W.B., Pedlar, A., Garrett, M.A., Wills, K.A., Garrington, S.T., Diamond, P.J. and Wilkinson, P.N.: 2001, *MNRAS* **322**, 100.

Miyoshi, M., Moran, J., Herrnstein, J., Greenhill, L., Nakai, N., Diamond, P. and Inoue, M.: 1995, *Nature* **373**, 127.

Momjian, E., Romney, J.D., Carilli, C.L. and Troland, T.H.: 2003, *ApJ* **597**, 809.

Mundell, C.G., Pedlar, A., Baum, S.A., O'Dea, C.P., Gallimore, J.F. and Brinks, E.: 1995, *MNRAS* **272**, 355.

Mundell, C.G., Pedlar, A., Shone, D.L. and Robinson, A.: 1999, *MNRAS* **304**, 481.

Mundell, C.G., Ferruit, P. and Pedlar, A.: 2001, *ApJ* **560**, 168.

Mundell, C.G., Wrobel, J.M., Pedlar, A. and Gallimore, J.F.: 2003, *ApJ* **583**, 192.

Muxlow, T.W.B., Pedlar, A., Wilkinson, P.N., Axon, D.J., Sanders, E.M. and de Bruyn, A.G.: 1994, *MNRAS* **266**, 455.

Neff, S.G. and Ulvestad, J.S.: 2000, *AJ* **120**, 670.

Pedlar, A., Meaburn, J., Axon, D.J., Unger, S.W., Whittle, D.M., Meurs, E.J.A., Guerrine, N. and Ward, M.J.: 1989, *MNRAS* **238**, 863.

Pedlar, A., Fernandez, B., Hamilton, N.G., Redman, M.P. and Dewdney, P.E.: 1998, *MNRAS* **300**, 1071.

Pedlar, A., Muxlow, T.W.B., Smith, R., Thrall, H., Beswick, R., Argo, M., Aalto, S., Booth, R.S. and Wills, K.A.: in press, in: *The Neutral ISM in Starburst Galaxies*, ASP Conference Series.

Pihlström, Y.M., Conway, J.E., Booth, R.S., Diamond, P.J. and Polatidis, A.G.: 2001, *A&A* **377**, 413.

Rees, M.J.: 1984, *ARA&A* **22**, 417.

Rovilos, E., Diamond, P.J., Lonsdale, C.J., Lonsdale, C.J. and Smith, H.E.: 2003, *MNRAS* **342**, 373.

Seaquist, E.R., Frayer, D.T. and Frail, D.A.: 1997, *ApJ* **487**, 131.

Shen, J. and Lo, K.Y.: 1995, *ApJ* **445**, 99.

Smith, H.E., Lonsdale, C.J., Lonsdale, C.J. and Diamond, P.J.: 1998, *ApJ* **493**, 17.

Warwick, R.S., Koyama, K., Inoue, H., Takano, S., Awaki, H. and Hoshi, R.: 1989, *PASJ* **41**, 739.

Weiss, A., Walter, F., Neininger, N. and Klein, U.: 1999, *A&A* **345**, 23.

Wills, K.A., Pedlar, A. and Muxlow, T.W.B.: 1998, *MNRAS* **298**, 347.

Wills, K.A., Pedlar, A., Muxlow, T.W.B. and Stevens, I.R.: 1999, *MNRAS* **305**, 680.

Wills, K.A., Das, M., Pedlar, A., Muxlow, T.W.B. and Robinson, T.G.: 2000, *MNRAS* **316**, 33.

Wills, K.A., Pedlar, A. and Muxlow, T.W.B.: 2002, *MNRAS* **331**, 313.

Yun, M.S. and Hibbard, J.E.: 2001, *ApJ* **550**, 104.

CO, ^{13}CO, AND [CI] IN GALAXY CENTERS

F.P. ISRAEL
Sterrewacht Leiden, NL, RA Leiden, The Netherlands; E-mail: israel@strw.leidenuniv.nl

(Received 16 April 2004; accepted 15 June 2004)

Abstract. Measurements of [CI], $J = 2$–1 ^{13}CO and $J = 4$–3 ^{12}CO emission from quiescent, starburst and active galaxy centers reveal a distinct pattern characterized by relatively strong [CI] emission. The [CI] to ^{13}co emission ratio increases with central [CI] luminosity. It is lowest in quiescent and mild starburst centers and highest for strong starburst centers and active nuclei. C° abundances are close to, or even exceed, CO abundances. The emission is characteristic of warm and dense gas rather than either hot tenuous or cold very dense gas. The relative intensities of CO, [CI], [CII] and far-infrared emission suggest that the dominant excitation mechanism in galaxy centers may be different from that in photon-dominated regions (PDRs).

Keywords: galaxies, ISM, molecules, carbon lines

1. Introduction

Carbon monoxide (CO), the most common molecule after H_2, is routinely detected in external galaxies. When exposed to photons of energy greater than 11.1 eV, CO is readily photodissociated into atomic carbon and oxygen. As the C° ionization potential is only 11.3 eV, i.e. quite close to the CO dissociation energy, neutral carbon may be quickly ionized. Because carbon monoxide, its isotopes, as well as neutral and ionized carbon respond differently to ambient conditions, observations of the relative emission strengths of ^{12}CO, ^{13}CO, C° and C^+ provide significant information on the physical condition of cloud complexes from which the emission arises. Even though far-infrared continuum and [CII] lines are much more efficient coolants overall, CO and [CI] lines are especially important for the temperature balance of cool and dense molecular gas.

2. Relative Intensities of CO, ^{13}CO, and [CI]

CO may be observed from the ground in many transitions. Emission from C^+ has been observed towards numerous galaxies from the KAO and ISO platforms. Emission from C° can also be measured from the ground but only under excellent atmospheric conditions. To date, the total number of galaxies detected in [CI] is about 30. Most of the extragalactic 492 GHz [CI] measurements are summarized in just two papers, by Gerin and Phillips (2000) using the CSO and by Israel and Baas (2002) using the JCMT, both on Mauna Kea, Hawaii. These papers include references to more detailed discussions of individual galaxy results.

Almost all galaxies mapped thus far show strong concentrations of both atomic carbon and molecular gas well contained within radii $R \leq 0.5$ *kpc*. The area-integrated CO and [CI] luminosities of the observed galaxy centers cover a large range. Quiescent centers (NGC 7331, IC 342, Maffei 2, NGC 278, NGC 5713) have modest [CI] luminosities $\approx 1 \leq L_{[CI]} \leq 5\,\mathrm{K\,km\,s^{-1}\,kpc^2}$ (with $1\,\mathrm{K\,km\,s^{-1}kpc^2}$ corresponding to 2.2×10^{20} W). Starburst nuclei (NGC 253, NGC 660, M 82, NGC 3628, NGC 6946) have luminosities $10 \leq L_{[CI]} \leq 40\,\mathrm{K\,km\,s^{-1}\,kpc^2}$, except M 83 which has only $L_{[CI]} = 3.6\,\mathrm{K\,km\,s^{-1}\,kpc^2}$. The highest luminosities, $L_{[CI]} \geq 50\,\mathrm{K\,km\,s^{-1}\,kpc^2}$, are found around the active nuclei of NGC 1068 and NGC 3079.

In previously observed Galactic photon-dominated regions (PDRs), the intensities of $^3P_1 - ^3P_0$ [CI] and $J = 2$–1 ^{13}CO line emission were generally found to be very similar (cf. Keene et al., 1996; Kaufman et al., 1999). Such ratios of $^3P_1 - ^3P_0$ [CI] and $J = 2$–1 ^{13}CO close to or less than unity are considered to be characteristic of the effects of enhanced UV radiation on molecular gas in PDRs. However, Figure 1 shows that galaxy centers may have much stronger [CI] emission.

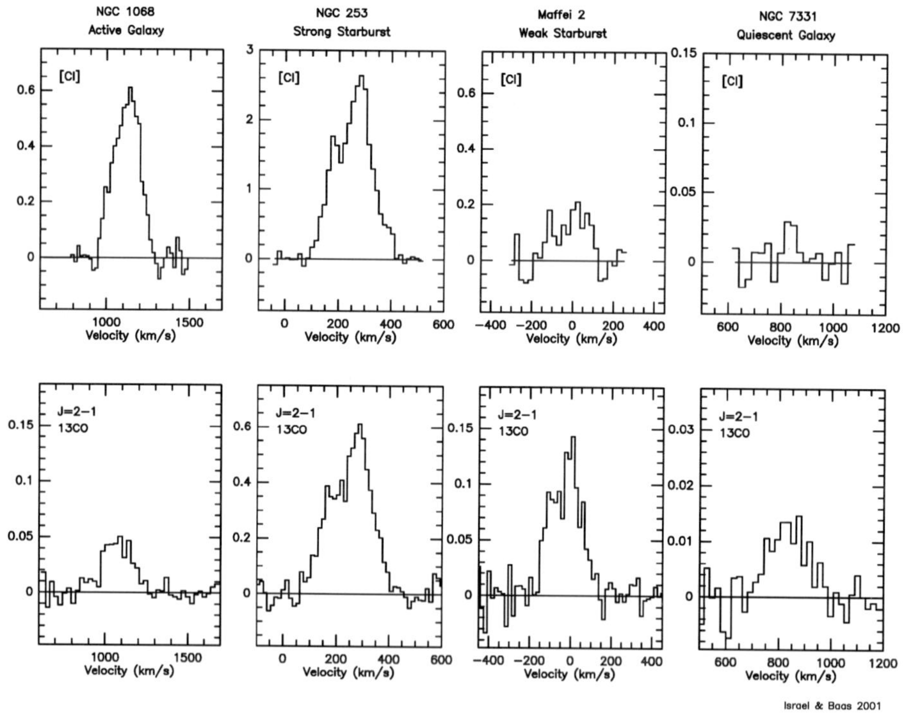

Figure 1. [CI] (top) and $J = 2$–1 ^{13}CO (bottom) spectra observed towards galaxy centers. The vertical scale is T_{mb} in K; the horizontal scale is velocity V_{LSR} in $\mathrm{km\,s^{-1}}$. Note decreasing ratio [CI]^{13}CO going from left to right.

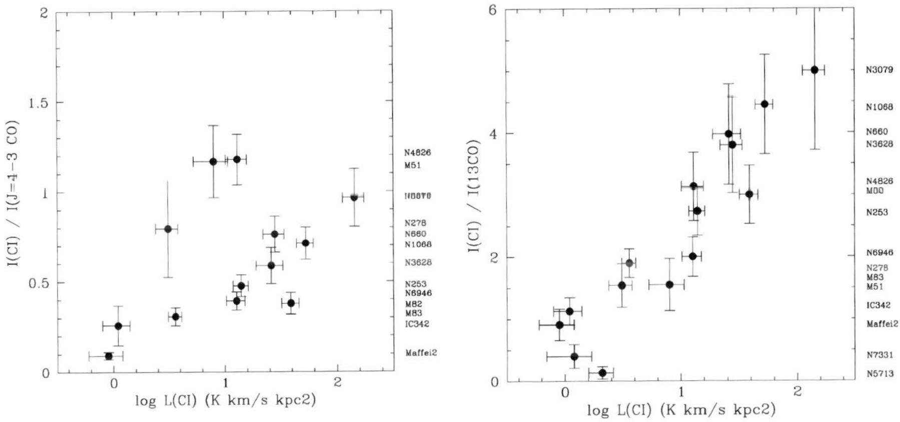

Figure 2. Left: [CI]/(J = 4–3 ^{12}CO) ratios versus area-integrated luminosity L[CI]. Right: [CI]/(J = 2–1 ^{13}CO) ratios versus L[CI]. The I([CI])/I(^{13}CO) ratio appears to be a well-defined function of log L([CI]); the I([CI])/I(4–3^{12}CO) ratio is not. Galactic sources (not shown) would all be crowded together in the lower left corner.

In Figure 2 we present a full comparison of the intensities of the 492 GHz [CI] line, the 461 GHz J = 4–3 ^{12}CO line and the 220 GHz J = 2–1 ^{13}CO line. The $^3P_1 - {}^3P_0$ [CI] line is stronger than J = 2–1 ^{13}CO in all except three galaxies. The highest [CI]/^{13}CO ratios of about five belong to the active galaxies NGC 1068 and NGC 3079. Generally, the $^3P_1 - {}^3P_0$ [CI] line is weaker than the J = 4–3 ^{12}CO line, but not by much. *Thus, galaxy centers have much stronger [*CI*] emission than the Galactic PDR results would lead us to expect.*

In the galaxy sample independently observed by Gerin and Phillips (2000) more than two-thirds also have a ratio [CI]/^{13}CO ≥ 2. Low ratios are expected in high-UV environments, and in environments with high gas (column) densities where virtually all carbon will be locked up in CO. In the Galaxy, high ratios are found in environments with low gas column densities and mild UV radiation fields, such as exemplified by translucent clouds and at cloud edges. There, ^{12}CO and especially ^{13}CO will be mostly dissociated, and gas-phase atomic carbon may remain neutral. In dense clouds, [CI] may be relatively strong if the dominant heating mechanism is some other than UV photons. The data presented here and in Gerin and Phillips (2000) suggest that most of *the emission from galaxy centers does not come from very dense, star-forming molecular cloud cores or PDR zones.*

3. Modeling of [CI], CO, and ^{13}CO

Very approximate column densities may be calculated assuming optically thin [CI] and ^{13}CO emission in the high-temperature LTE limit, but accurate results are obtained only by more detailed radiative transfer calculations. Curves illustrating the

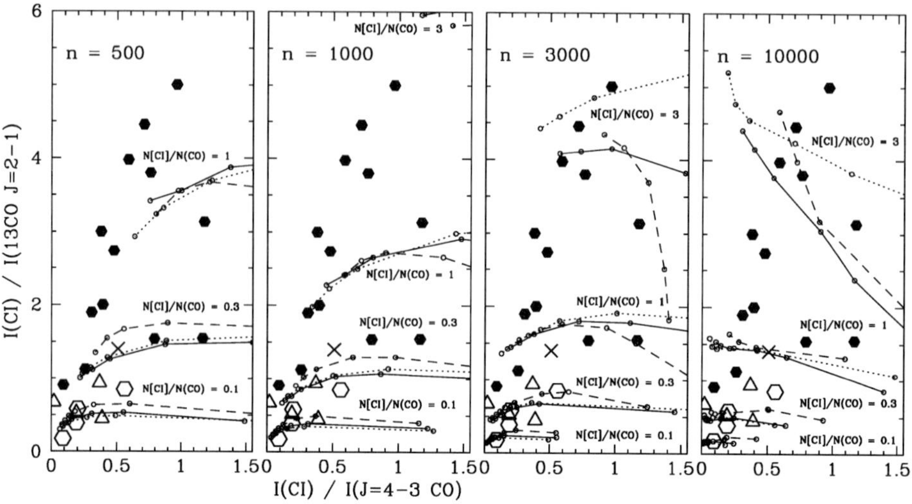

Figure 3. Observed line intensity ratios [CI]/^{13}CO versus [CI]/CO 4–3 compared to LVG radiative-transfer model ratios at selected gas densities, temperatures and velocity gradients assuming an isotopic ratio [^{12}CO]/[^{13}CO] = 40. Galaxy centers are marked by filled hexagons, PDRs by open hexagons (LMC) or open triangles (Milky Way), and the Milky Way Center by a cross. Line families indicate abundances N[CI]/N(CO) = 0.1, 0.3, 1 and 3. Within each family, a dotted line corresponds to a gradient $N(\mathrm{CO})/dV = 3 \times 10^{16}\,\mathrm{cm}^{-2}/\,\mathrm{km\,s}^{-1}$, solid line to $N(\mathrm{CO})/dV = 1 \times 10^{17}\,\mathrm{cm}^{-2}/\,\mathrm{km\,s}^{-1}$ and dashed lines to $N(\mathrm{CO})/dV = 3 \times 10^{17}\,\mathrm{cm}^{-2}/\,\mathrm{km\,s}^{-1}$. On each track, temperatures of 150, 100, 60, 30, 20 and 10 K are marked from left to right by small open circles.

possible physical condition of gas clouds, shown in Figure 3, were calculated with the Leiden radiative transfer models (see http://www.strw.leidenuniv.nl/~radtrans/). The four panels correspond to molecular gas densities ranging from $n(\mathrm{H_2}) = 500\,\mathrm{cm}^{-3}$ to $n(\mathrm{H_2}) = 10000\,\mathrm{cm}^{-3}$. In each panel, mostly horizontal tracks mark CO gradients $N(\mathrm{CO})/dV$ =0.3, 1.0 and 3.0 $\times 10^{17}$ $\mathrm{cm}^{-2}/\,\mathrm{km\,s}^{-1}$ for [C°]/CO abundances of 0.1, 0.3, 1.0 and 3.0, respectively, and kinetic temperatures ranging from 150 K (left) to 10 K (right) are marked. Lines of constant kinetic temperature would be mostly vertical in these panels. The diagrams contain points representing the observed galaxy centers as well as star-forming regions (White and Sandell, 1995; Israel and Baas, unpublished; Bolatto et al., 2000) and the Milky Way Center (Fixsen et al., 1999).

The star-forming PDRs have relatively low neutral carbon versus CO abundances (C°)/CO) $\approx$ 0.1–0.3). *Most galaxy centers (especially the active ones) have significantly higher abundances exceeding unity*, independent of the assumed gas parameters. The diagonal distribution of galaxy centers roughly follows lines of constant kinetic temperature. The actual temperatures T_{k} and molecular gas densities $n(\mathrm{H_2})$ cannot be determined independently. For instance, if we assume $n = 500\,\mathrm{cm}^{-3}$, we find $T_{\mathrm{kin}} > 150$, whereas by assuming $n \geq 3000\,\mathrm{cm}^{-3}$ we find more modest values $T_{\mathrm{kin}} = 30 - 60$ K similar to the dust temperatures 33 K $\leq T_{\mathrm{d}} \leq$ 52 K found for these galaxy centers by Smith and Harvey (1996). The only available direct determination, for M 82 by Stutzki et al. (1997), yields $n \geq 10^4\,\mathrm{cm}^{-3}$ and $T_\mathrm{k} = 50$–100 K, in very good agreement with the above. Where the molecular line emission in most galaxy centers appears to arise from warm, dense gas (as opposed to either hot and tenuous or cold and very dense gas), the centers of NGC 7331, M 51 and NGC 4826 seem to be cooler independent of assumed gas density.

4. [CI], [CII], and FIR Intensities

Figure 4 combines [CI] line intensities with those of [CII] and the far-infrared continuum (for references, see Israel and Baas, 2002). The results resemble those obtained by Gerin and Phillips (2000). There is no longer a clear distinction between the various types of objects. The [CII]/FIR ratio increases with [CI]/FIR, but the [CII]/[CI] ratio *decreases* as [CI]/FIR increases. An upper limit for [CII] places the merger galaxy NGC 660 in the same diagram positions as the ultraluminous mergers Arp 220 and Mrk 231 observed by Gerin and Phillips. The highest [CII]/FIR ratios are reached for PDR model gas densities $n = 10^3 - 10^4\,\mathrm{cm}^{-3}$ (Gerin and Phillips, 2000), but fully half of the ratios in Figure 4 are much higher than predicted by

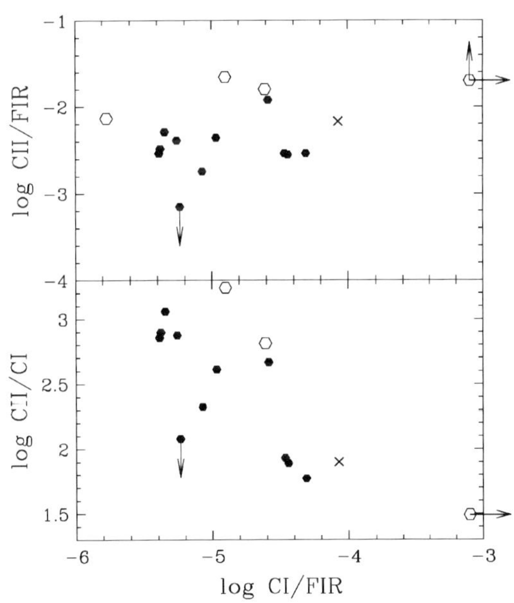

Figure 4. Ionized carbon [CII] line to far-infrared continuum (FIR) ratio (top) and [CII]/[CI] line ratio (bottom) as a function of neutral carbon [CI] line to FIR ratio. In both diagrams, the position of the Milky Way Center is marked by a cross and the positions of Magellanic Cloud objects by open hexagons.

The differences in luminosity and ionizing flux between low-mass and high-mass objects causes large variations in the chemistry of the surrounding material, although the basic processes are the same.

The chemistry in disks also follows the same processes, but densities are higher, and even weak ionizing fluxes from the central star are important because of the small distances.

Much of what we know about this chemistry has been learnt with single-dish telescopes such as CSO and JCMT; and interferometer arrays such as PdBI, OVRO, and BIMA. In the coming years, our understanding will greatly benefit from observations with APEX, the SMA and its link-up with JCMT and CSO, the merger of OVRO and BIMA into CARMA, and ultimately ALMA. These instruments will be complemented at shorter wavelengths by the Spitzer, Herschel, and SOFIA.

Together these observational possibilities will allow access to higher excitations, tracing warmer and denser material; and higher spatial resolutions, allowing to separate the different regions that are still confused around high-mass stars and in protoplanetary disks. Paired with increased modeling capabilities, the foreseen observations will greatly increase our understanding of the chemical evolution of protostars.

References

Bachiller, R.: 1996, *ARA&A* **34**, 111.

Bacmann, A., Lefloch, B., Ceccarelli, C., Steinacker, J., Castets, A. and Loinard, L.: 2003, *ApJ* **585**, L55.

Beckwith, S.V.W. and Sargent, A.I.: 1996, *Nature* **383**, 139.

Bergin, E.A., Neufeld, D.A. and Melnick, G.J.: 1998, *ApJ* **499**, 777.

Bergin, E.A., Alves, J., Huard, T. and Lada, C.J.: 2002, *ApJ* **570**, L101.

Bergin, E., Calvet, N., D'Alessio, P. and Herczeg, G.J.: 2003, *ApJ* **591**, L159.

Bergin, E.A., Hartmann, L.W., Raymond, J.C. and Ballesteros-Paredes, J.: in press, *ApJ* **612**, 921.

Caselli, P., van der Tak, F.F.S., Ceccarelli, C. and Bacmann, A.: 2003, *A&A* **403**, L37.

Ceccarelli, C., Dominik, C., Lefloch, B., Caselli, P. and Caux, E.: 2004, *ApJ* **607**, L51.

Churchwell, E.: 2002, *ARA&A* **40**, 27.

Dalgarno, A.: 1987, in: G. Morfill and M.S. Schiler (eds.), *NATO ASIC Proceedings of the 210: Physical Processes in Interstellar Clouds*, Reidel, Dordrecht, p. 219.

Draine, B.T.: 1978, *ApJS* **36**, 595.

Dutrey, A., Guilloteau, S. and Guelin, M.: 1997, *A&A* **317**, L55.

Evans, N.J.: 1999, *ARA&A* **37**, 311.

Hartmann, L., Ballesteros-Paredes, J. and Bergin, E.A.: 2001, *ApJ* **562**, 852.

Jørgensen, J.K., Hogerheijde, M.R., Blake, G.A., van Dishoeck, E.F., Mundy, L.G. and Schöier, F.L.: 2004, *A&A* **415**, 1021.

Jørgensen, J.K., Schöier, F.L. and van Dishoeck, E.F.: 2004, *A&A* **416**, 603.

Kuiper, T.B.H., Langer, W.D. and Velusamy, T.: 1996, *ApJ* **468**, 761.

Kurtz, S., Cesaroni, R., Churchwell, E., Hofner, P. and Walmsley, C.M.: 2000, in: V. Mannings, A.P. Boss and S.S. Russell (eds.), *Protostars and Planets IV*, Tucson, University of Arizona Press, p. 299.

Langer, W.D., van Dishoeck, E.F., Bergin, E.A., Blake, G.A., Tielens, A.G.G.M., Velusamy, T. and Whittet, D.C.B.: 2000, in: V. Mannings, A.P. Boss and S.S. Russell (eds.), *Protostars and Planets IV*, Tucson, University of Arizona Press, p. 29.

Le Teuff, Y.H., Millar, T.J. and Markwick, A.J.: 2000, *A&AS* **146**, 157.

Lis, D.C., Roueff, E., Gerin, M., Phillips, T.G., Coudert, L.H., van der Tak, F.F.S. and Schilke, P.: 2002, *ApJ* **571**, L55.

Lucas, R. and Liszt, H.S.: 1997, in: E.F. van Dishoeck (ed.), *Proceedings of the IAU Symposium 178, Molecules in Astrophysics: Probes and Processes*, Kluwer Academic Publishers, Dordrecht, p. 421.

Meyer, D.M.: 1997, in: E.F. van Dishoeck (ed.), *Proceedings of the IAU Symposium 178, Molecules in Astrophysics: Probes and Processes*, Kluwer Academic Publishers, Dordrecht, p. 407.

Olano, C.A., Walmsley, C.M. and Wilson, T.L.: 1988, *A&A* **196**, 194.

Parise, B., Castets, A., Herbst, E., Caux, E., Ceccarelli, C., Mukhopadhyay, I. and Tielens, A.G.G.M.: 2004, *A&A* **416**, 159.

Qi, C.: 2001, PhD Thesis, California Institute of Technology.

Ruffle, D.P., Rae, J.G.L., Pilling, M.J., Hartquist, T.W. and Herbst, E.: 2002, *A&A* **381**, L13.

Schilke, P., Walmsley, C.M., Pineau des Forets, G. and Flower, D.R.: 1997, *A&A* **321**, 293.

Schutte, W.A.: 1997, in: J.M. Greenberg (ed.), *NATO ASIC Proceedings of the 487: The Cosmic Dust Connection*, Kluwer Academic Publishers, Dordrecht, p. 177.

Semenov, D., Wiebe, D. and Henning, T.: 2004, *A&A* **417**, 93.

van der Tak, F.F.S., van Dishoeck, E.F. and Caselli, P.: 2000, *A&A* **361**, 327.

van der Tak, F.F.S., Schilke, P., Müller, H.S.P., Lis, D.C., Phillips, T.G., Gerin, M. and Roueff, E.: 2002, *A&A* **388**, L53.

van Dishoeck, E.F.: 1988, in: R.D. Wolstencroft and W.B. Burton (eds.), *Millimetre and Submillimetre Astronomy*, Kluwer Academic Publishers, Dordrecht, p. 117.

van Dishoeck, E.F. and Blake, G.A.: 1998, *ARA&A* **36**, 317.

van Dishoeck, E.F. and Hogerheijde, M.R.: 1999, in: C.J. Lada and N.D. Kylafis (eds.), *NATO ASIC Proceedings of the 540: The Origin of Stars and Planetary Systems*, Kluwer Academic Publishers, Dordrecht, p. 97.

van Zadelhoff, G.-J., van Dishoeck, E.F., Thi, W.-F. and Blake, G.A.: 2001, *A&A* **377**, 566.

Wiebe, D., Semenov, D. and Henning, T.: 2003, *A&A* **399**, 197.

Willacy, K., Langer, W.D. and Velusamy, T.: 1998, *ApJ* **507**, L171.

Williams, D.A.: 1993, in: T.J. Miller and D.A. Williams (eds.), *Dust and Chemistry in Astronomy*, Institute of Physics Publishing, Philadelphia, p. 143.

MODEL OF THE W3(OH) ENVIRONMENT BASED ON DATA FOR BOTH MASER AND "QUASI-THERMAL" METHANOL LINES

ANDREI M. SOBOLEV[1], EDMUND C. SUTTON[2], DINAH M. CRAGG[3] and PETER D. GODFREY[3]

[1] *Ural State University, Russia*
[2] *University of Illinois at Urbana-Champaign, USA*
[3] *Monash University, Australia*

(Received 16 April 2004; accepted 15 June 2004)

Abstract. In studies of the environment of massive young stellar objects, recent progress in both observations and theory allows a unified treatment of data for maser and "quasi-thermal" lines. Interferometric maser images provide information on the distribution and kinematics of masing gas on small spatial scales. Observations of multiple masing transitions provide constraints on the physical parameters.

Interferometric data on "quasi-thermal" molecular lines permits an investigation of the overall distribution and kinematics of the molecular gas in the vicinity of young stellar objects, including those which are deeply embedded. Using multiple transitions of different molecules, one can obtain good constraints on the physical and chemical parameters. Combining these data enables the construction of unified models, which take into account spatial scales differing by orders of magnitude.

Here, we present such a combined analysis of the environment around the ultracompact HII region in W3(OH). This includes the structure of the methanol masing region, physical structure of the near vicinity of W3(OH), detection of new masers in the large-scale shock front and embedded sources in the vicinity of the TW young stellar object.

Keywords: radiolines, ISM: molecules

1. Possibility of Combined Analysis of Data on Maser and "Quasi-Thermal" Lines

Modern studies of the kinematical, physical and chemical status of star-forming regions are largely based on the information delivered by emission in molecular lines. Different molecular transitions have different excitation temperatures, as a result of the radiative and collisional population exchange between quantum levels. In the stationary case these take the form of pumping cycles (see, e.g., Sobolev and Deguchi, 1994). Excitation states of molecular transitions are mostly characterized by positive excitation temperatures and, hence, the optical depths of the corresponding lines are mostly positive. Such lines are called "quasi-thermal" and trace the whole extent of the sufficiently dense and warm environment of the massive young stellar object where the corresponding molecule is abundant.

Astrophysics and Space Science **295:** 189–196, 2004.

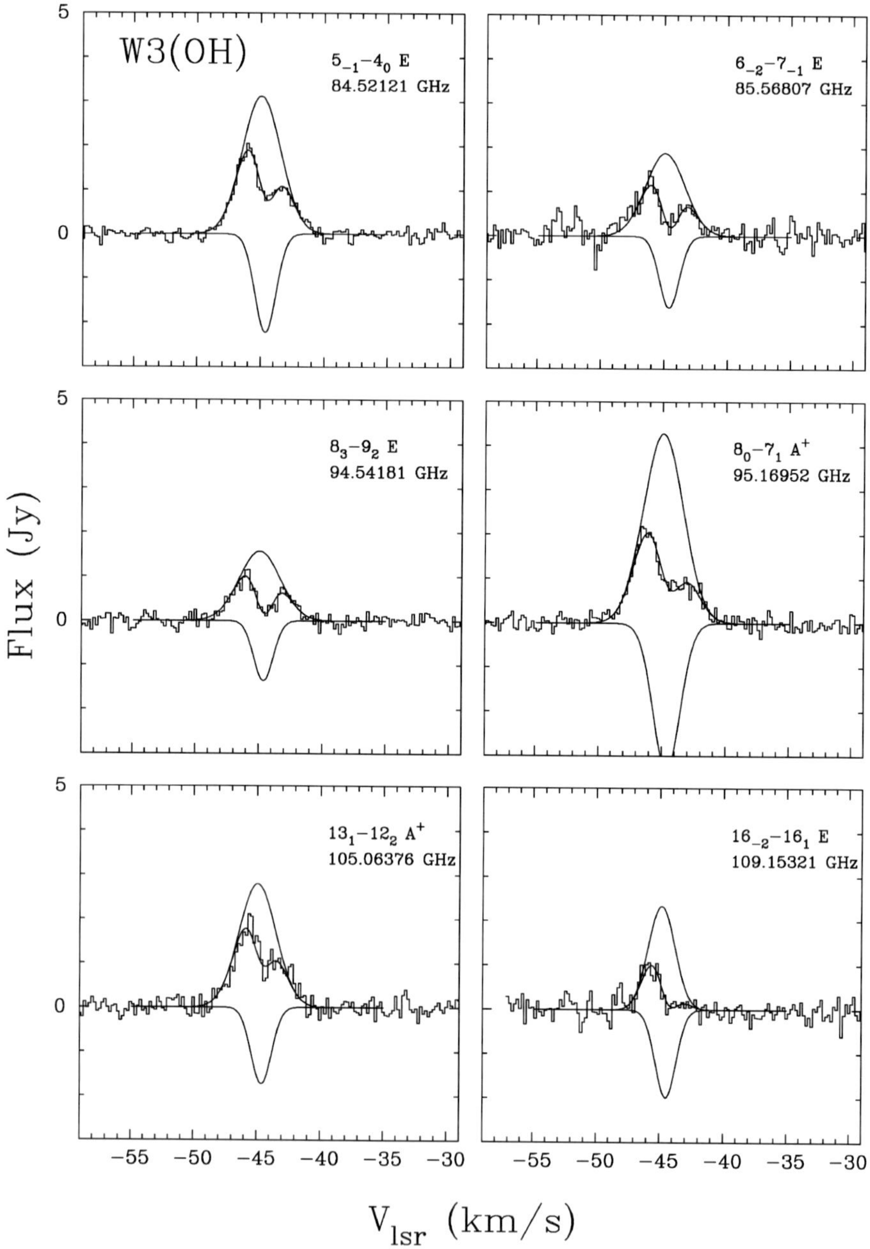

Figure 1. BIMA spectra of six methanol lines which were predicted to absorb in the methanol masing region. Fits with two gaussians corresponding to common emitting ($V_{\rm LSR} = -44.9$ km/s) and absorbing ($V_{\rm LSR} = -44.6$ km/s) regions are shown. The line profiles are not consistent with two emission components at fixed velocities. This suggests that there is an absorbing region situated closer to the observer. So, it is likely that we see absorption against the line emitting background.

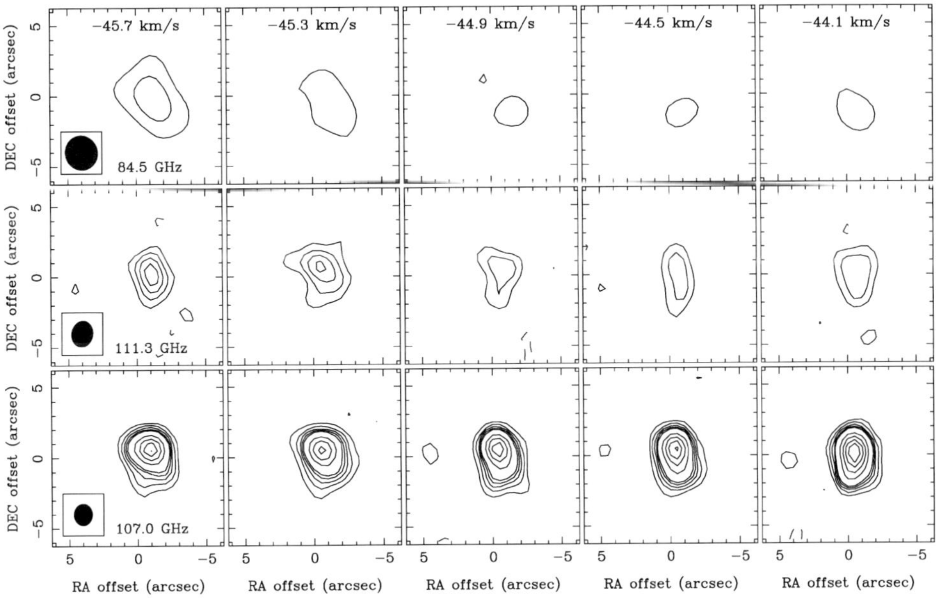

Figure 2. BIMA maps of W3(OH) in the methanol lines. The upper row of channel maps corresponds to the line at 84.5 GHz which shows strong absorption in the model (Sutton et al., 2001) and displays an absorption dip in the spectrum shown in Figure 1. It is seen that no emission comes from the northern part of the source in the central and right panels corresponding to velocities of the dip. The middle row shows maps of the 111.3 GHz line which is transparent in the model and shows an uneven pedestal profile. It is seen that emission comes from both northern and southern parts of the source at all line velocities. The lower row of panels shows maps of the 107.0 GHz line which is observed as a bright maser and behaves accordingly in the model. It is seen that emission in the central and right panels mostly comes from the northern part of the source where the methanol masing region described in the text is situated. Contour levels are 0.4, 0.8, 1.2, 1.6, 2, 4, 6, 8, and 10 Jy/beam. Interferometer beams are shown in the lower left corners of the leftmost panels.

the UCHII, and the second region producing the strong masers and absorptions. The latter has a size of order 1 arcsecond, which is about 3 orders of magnitude greater than that of the 6.7 GHz maser spots, and corresponds to the total extent of the region where the strong class II methanol masers are distributed. So, BIMA observations have shown that the methanol maser region in W3(OH) has a size of a few by 10^{16} cm, and provided estimates of the physical parameters on the basis of combined analysis of the masing and quasi-thermally excited lines.

3. How "Quasi-Thermal" Line Observations Help Reveal Masers and Vice Versa

Sutton et al. (2004) showed that BIMA observations reveal the presence of extended features to the south-west of W3(OH) which are traced by the methanol line emission (see also lower panels in Figure 2). Methanol is a chemical tracer of shocked regions, and analysis of the spatial structure of its emission in regions of massive star formation can lead to the discovery of extremely young stellar objects displaying maser emission (Sobolev and Strelnitskii, 1982). Indeed, Sutton et al. (2004) found two objects which are most pronounced in the Class I methanol maser lines. Excitation analysis carried out in this paper shows that these lines are most probably weak masers, which can be confirmed by interferometric measurements with higher spatial resolution. This can be done with the forthcoming CARMA facility or with the VLA in counterpart lines of the same methanol line series.

In turn, data on the maser emission of water around the TW young stellar object situated about 6″ to the east of W3(OH) shows the presence of molecular material situated on both sides of the TW object with an almost equal angular separation of about 1". The nature of the material in the immediate vicinity of the TW object is unclear and can be explained in terms of the jet and edge-on disk (Shchekinov and Sobolev, 2004). Water masers are produced under very special conditions and can display great deviations from the systemic velocity of the source, i.e., they might not reveal bulk motions of matter. So, observations of "quasi-thermal" molecular emission in the TW vicinity with appropriate angular resolution are very important to clarify the situation and, in any case, help promote understanding of important

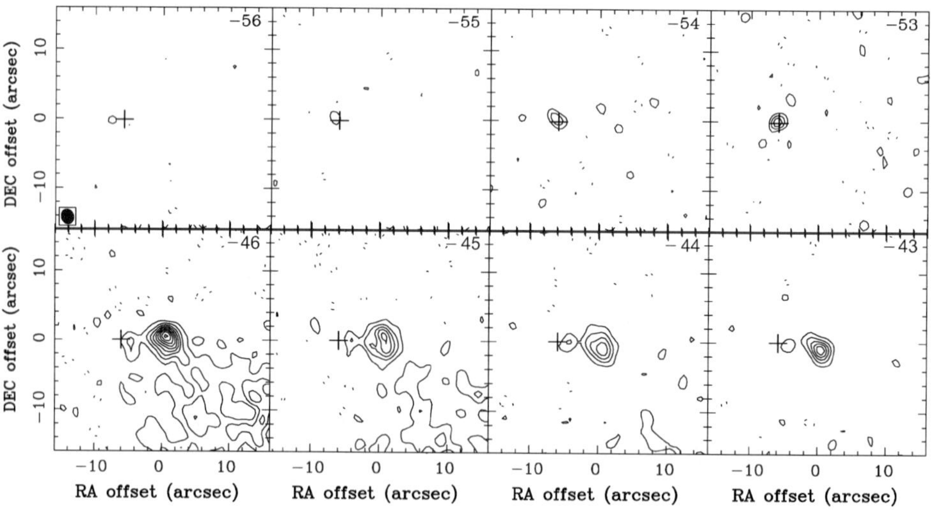

Figure 3. BIMA channel maps of the TW vicinity in the methanol lines at 96.7 GHz. The cross marks the position of the TW object. The upper row of panels shows emission in the blue wing of the 96.74458 GHz line and the leftmost upper panel displays embedded material situated about 1″ east of the TW. The lower row of panels corresponds to the red wing of the 96.73939 GHz line and displays the embedded source to the west of the TW. The interferometer beam is shown in the lower left corner of the leftmost upper panel.

processes related to the earliest stages of massive star formation. The existence of molecular material situated 1" to the west of TW was clearly shown by interferometry of red-shifted highly excited "quasi-thermal" line emission (Wyrowski et al., 1997) and a compact continuum source was found at this position (Wyrowski et al., 1999). However, highly excited lines and the continuum did not indicate the presence of molecular material to the east of the TW object where the strongest water masers reside.

In the paper by Sutton et al. (2004) it is shown that the TW object is surrounded by a dense envelope of molecular gas. This surrounding gas obscures the source interiors. However, Doppler shifted internal objects can display themselves in the line wings. We have searched for such emission in the wings of the strong low excitation lines of the methanol line quartet at 96.7 GHz. Because the lines are blended, we examined the blue wing of the 96.74458 GHz line and the red wing of the 96.73939 GHz line. Results of the search are shown in Figure 3. We found the source to the west of TW in the red wing emission and found evidence of the existence of the blue-shifted methanol emission from the water maser site situated about 1" east of TW. The latter fact is also interesting from the chemical point of view because it shows that the water masers are formed in the methanol-abundant region. So, masers have prompted the detection of "quasi-thermal" line emission which can provide a clue for elucidating the nature of one of the most intriguing objects of early massive star formation.

4. Conclusions

This paper provides an example of how combined analysis of the interferometry data on the maser and "quasi-thermal" lines helps to elucidate the structural, physical and chemical status of the regions of massive star formation. Data on sets of both maser and "quasi-thermal" lines allows estimates of the temperature, density, molecular abundances and source extensions and shapes. We find that the numerous methanol transitions form a good basis for such types of studies both from the point of view of their high diagnostic capacity and because relevant lines are rather easy to observe with existing and forthcoming radioastronomical facilities. We have also shown that information on "quasi-thermal" lines can reveal the existence of masing objects while the masers can prompt the detection of molecular material emitting in "quasi-thermal" lines.

Acknowledgements

AMS was supported by grants from RFBR (03-02-16433) and the Russian Ministry of Education (E02-11.0-43). ECS was supported by NSF Grant AST 02-28953. DMC and PDG were supported by the Australian Research Council.

References

Alcolea, J., Menten, K.M., Moran, J.M. and Reid, M.J.: 1993, The proper motions of the H2O masers near W3(OH). in: Astrophysical Masers; Proceedings of the Conference, Arlington, VA, Mar. 9–11, 1992, **259**, 225.

Gwinn, C.: 1994, *ApJ* **431**, L123.

Menten, K.M., Walmsley, C.M., Henkel, C., Wilson, T.L., Snyder, L.E., Hollis, J.M. and Lovas, F.J.: 1986, *A&A* **169**, 271.

Minier, V., Booth, R. and Conway, J.: 2002, *A&A* **383**, 614.

Shchekinov, Yu.A. and Sobolev, A.M.: 2004, *A&A* **418**, 1045.

Slysh, V.I., Kalenskii, S.V., Valtts I.E., Golubev V.V., and Mead K.: 1999, *ApJS* **123**, 515.

Sobolev, A.M.: 1993, *Astronomy Letters* **19**, 293.

Sobolev, A.M., Cragg, D.M. and Godfrey, P.D.: 1997, *MNRAS* **288**, L39.

Sobolev, A.M., Deguchi, S.: 1994, *ApJ* 433, 719.

Sobolev, A.M. and Strelnitskii, V.S.: 1983, *Soviet Astron. Letters* **9**, 12–16.

Sobolev, A.M., Wallin, B.K., Watson, W.D.: 1998, *ApJ* **478**, 763.

Sutton, E.C., Sobolev, A.M., Ellingsen, S.P., Cragg, D.M., Mehringer, D.M., Ostrovskii, A.B. and Godfrey, P.D.: 2001, *ApJ* **554**, 173.

Sutton, E.C., Sobolev, A.M., Salii, S.V., Malyshev, A.V., Ostrovskii, A.B. and Zinchenko, I.I.: 2004, *ApJ* **609**, 231.

Wilson, T.L., Johnston, K.J. and Mauersberger, R.: 1991, *A&A* **251**, 220.

Wyrowski, F., Hofner, P., Schilke, P., Walmsley, C.M., Wilner, D.J., and Wink, J.E.: 1997, *A&A* **320**, L17.

Wyrowski, F., Schilke, P., Walmsley, C.M. and Menten, K.M.: 1999, *ApJ* **514**, L43.

LONG-TERM MONITORING OF 6.7- AND 12.2-GHZ METHANOL MASERS

S. GOEDHART[1], M.J. GAYLARD[1] and D.J. VAN DER WALT[2]

[1] *Hartebeesthoek Radio Astronomy Observatory, Krugersdorp, South Africa; E-mail: sharmila@hartrao.ac.za*

[2] *Space Research Unit, Physics Department, NorthWest University, Potchefstroom, South Africa*

(Received 16 April 2004; accepted 15 June 2004)

Abstract. A sample of 54 6.7-GHz methanol maser sources was monitored at HartRAO for 4 years, and 11 12.2-GHz methanol masers for 3 years. The majority of the maser features display a significant degree of variability but with a wide range of timescales and behaviors. Some maser features remained unvarying throughout the monitoring programme, while others showed sporadic flares or sudden drops in flux density. Yet another group show quasi-periodic and periodic variations. In some cases the maser features dropped below the detection limit for a significant length of time before increasing in intensity and reappearing.

Keywords: masers, star formation

1. Introduction

Class II methanol masers are believed to be associated with the earliest stages of high mass star formation. These masers are excellent probes of changing conditions in the circumstellar environment. Survey observations of 6.7- and 12.2-GHz masers indicate that they are significantly variable (MacLeod and Gaylard, 1993, 1996; Caswell et al., 1995), but to date, large scale systematic studies of the nature of the variability have not been reported.

We can classify variability we might observe on a 4 year timescale into several groups: unchanging, monotonic increase or decrease, and fluctuations that are aperiodic, quasi-periodic or periodic. On longer timescales unchanging and monotonically increasing masers must of course change their behavior at some point since their environment is not stable and they will eventually vanish.

We can then consider possible causes for fluctuations. Aperiodic variations could be due to, for example, turbulence in the masing medium, in which case variations in different maser features would not be correlated; flares in the protostar due to accretion of clumps (variations are then likely to be correlated, with phase lags relating to light travel time); or clumpy outflows (phase lags may be seen between spatially separated maser spots on timescales related to clump velocities). Quasi-periodic variations require a resonance in the system, e.g., gravitational

Astrophysics and Space Science **295:** 197–202, 2004.

instability, perhaps a density wave, in an accretion disk. Periodic variations strongly constrain the underlying mechanism: pulsating star, starspots, rotating beam from protostar hotspot, precessing jets, binary or multiple star system, or orbiting protoplanets might all be possible. In each case we have to consider what range of periods is possible for each mechanism and what the plausibility of the mechanism might be.

Different scenarios regarding the maser location, pumping mechanism and the origin of variability can be further constrained by examining the time series for the following conditions:

- Are variations in different peaks correlated?
- Is there a time delay in variations between different peaks?
- What waveforms do variations exhibit?
- What is the behavior in different maser transitions?
- Is variability due to changes in existing maser spots, or do new spots appear and disappear?
- Do spot velocities change with time?
- What proper motions do the maser spots exhibit?

The location of the maser spots will impact on the mechanisms for variation that are possible. The masers could be located:

- At random in the cloud around the protostar
- In an infalling shell
- In a rotating circumstellar disk
- Entrained in outflowing molecular gas
- In clumps behind a shock front.

2. Observations

The observations were carried out using the Hartebeesthoek Radio Astronomy Observatory (HartRAO) 26-m telescope. The sources were observed at intervals of 1–2 weeks, with observations at 2–3 day intervals when a source was seen to be varying rapidly. The 6.7-GHz sample consisted of 54 sources which were observed from January 1999 to March 2003. A subset of 11 12.2-GHz sources were monitored during January 2000–March 2003.

3. Results

Sample data are shown to illustrate aspects of the range of behavior seen.

G59.78+0.06 (Figure 1) displayed a monotonic increase in brightness at 6.7 GHz. Conversely, G359.61−0.24 (Figure 2) showed an overall decrease in flux density, coupled with occasional low-amplitude fluctuations.

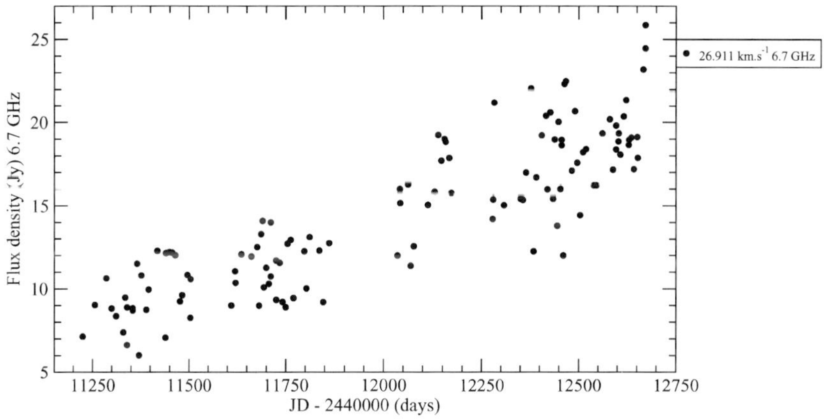

Figure 1. Time-series for a maser feature in G59.78+0.06. The legend gives the velocity of the mase feature and the transition.

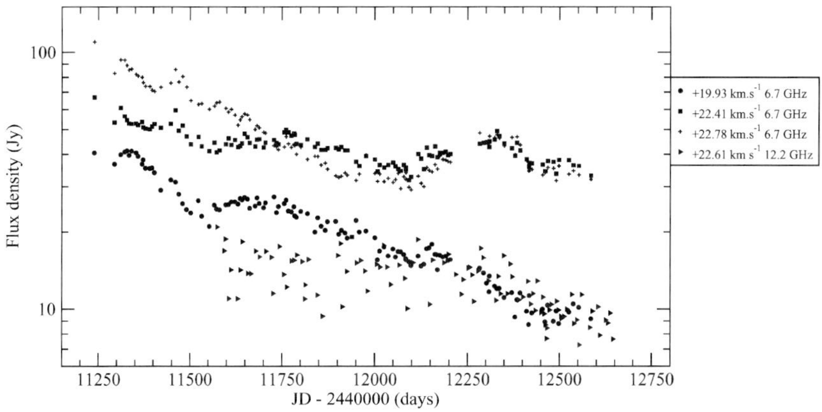

Figure 2. Time-series for selected maser features in G359.61−0.24.

G351.78−0.54 (Figure 3) has been monitored at HartRAO since 1992. From 1992 to 1998 the masers exhibited a steady decrease in flux density while undergoing short duration, aperiodic flares. Monitoring from 1999 showed that the long-term trend had reversed, while the sporadic flares continued to occur. These continue to display variable phase lags across the one maser peak (MacLeod and Gaylard, 1996).

G322.16+0.64 (Figure 4) is an example of a maser undergoing quasi-periodic variations. The intervals between maxima are 176 days, with a standard deviation of 30 days.

G9.62+0.20 (Figure 5) undergoes periodic flares at both 6.7- and 12.2 GHz, and in this case it is the 12.2-GHz masers that display a much larger amplitude change (Goedhart et al., 2003).

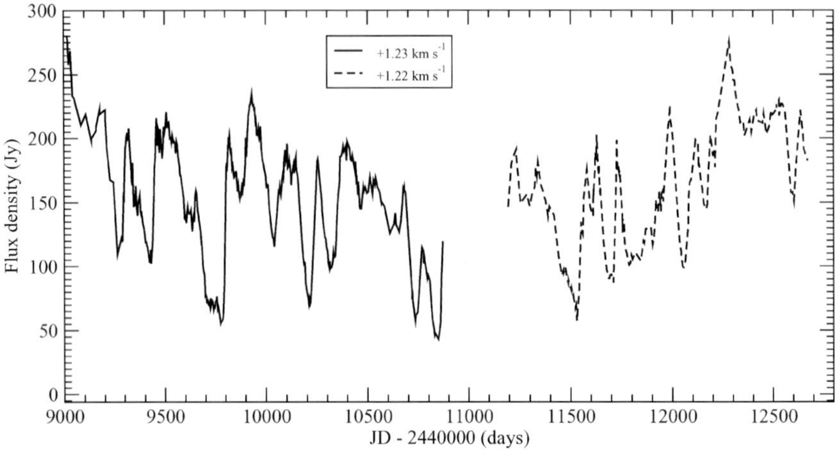

Figure 3. Time-series for a maser feature in G351.78−0.54.

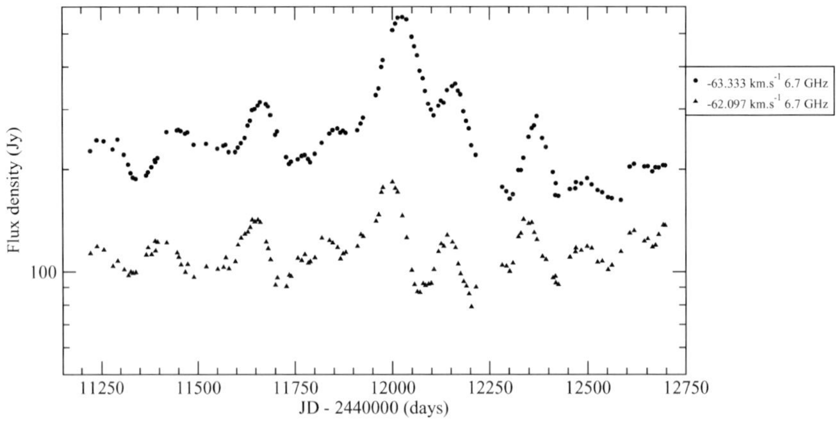

Figure 4. Time-series for selected maser features in G322.16+0.64.

G12.89+0.49 (Figure 6) shows rapid variations at both 6.7 and 12.2 GHz in the brightest maser feature, while a weaker maser feature at 6.7 GHz exhibits a much smaller range of amplitude, suggesting that the rapid variations are not instrumental in origin. The timescale of variation for this source is evidently less than 1 week, and is not resolved with our sampling interval.

A number of sources show flares happening at regular intervals, with periods ranging from 132 days up to a possible 668 days (Goedhart et al., in preparation). Eight sources show period uncertainties of just a few days, and we regard these sources as most likely periodic. Sources with a wider scatter in the times of the flares are classified as quasi-periodic. The range of periods seen strongly constrain the possible explanations, for example,

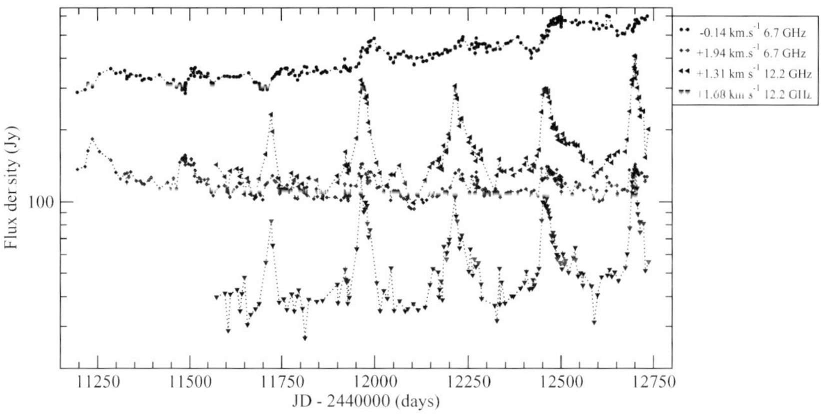

Figure 5. Time-series for selected maser features in G9.62+0.20.

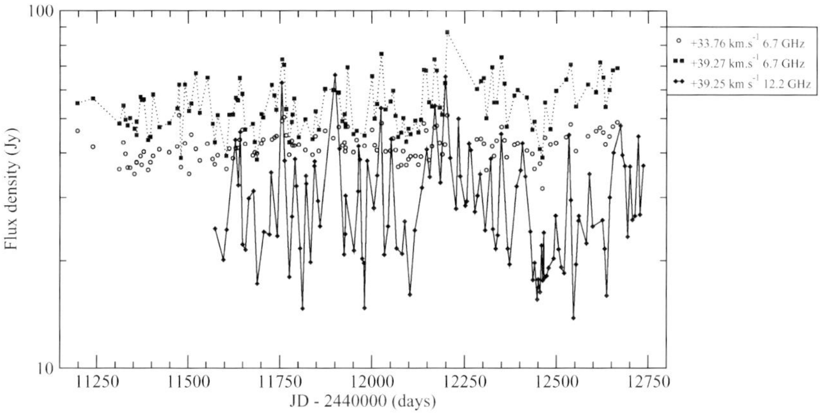

Figure 6. Time-series for selected maser features in G12.89+0.49.

- Pulsating stars are unlikely since β Cephei periods are <1 day (Kholopov et al., 1995)
- Hot or cold starspots seem unlikely as OB stars are fast rotators (Lang, 1992)
- Precessing jets are possible (Ouyed et al., 2003)
- Binary stars are possible (Bonnell and Bate, 2002)
- Gravitational instability in an accretion disk is possible as an explanation for quasi-periodicity (Durisen, 2001).

4. Summary and Conclusions

All the possible types of amplitude variation discussed in the introduction were seen in this large sample. Of greatest interest perhaps are the previously unknown periodically varying methanol masers, which show a wide range of periods. The timescales of general variations range from less than one week to greater than ten years. With two exceptions, the peak velocities of the maser features did not change, in contrast to the behavior often seen in water masers. The variations at 12.2 and 6.7 GHz were not necessarily correlated.

References

Caswell, J.L., Vaile, R.A. and Ellingsen, S.P.: 1995, *Publ. Astron. Soc. Aust.* **12**, 37.
Durisen, R.H., Mejia, A.C., Pickett, B.K. and Hartquist, T.W.: 2001, *ApJ* **563**, L157.
Goedhart, S., Gaylard, M.J. and van der Walt, D.J.: 2003, *MNRAS* **339**, L33.
Kholopov, P.N., Samus, N.N., Frolov, M.S., Goranskij, V.P., Gorynya, N.A., Kireeva, N.N., Kukarkina, N.P., Kurochkin, N.E., Medvedeva, G.I., Perova, N.B. and Shugarov, S.Y.: 1985, General Catalogue of Variable Stars (4th edn.), Nauka Publishing House, Moscow.
Lang, K.R.: 1992, Astrophysical Data: Planets and stars. Springer-Verlag.
MacLeod, G.C. and Gaylard, M.J.: 1993, in: A.W. Clegg and G.E. Nedoluha (eds.), Astrophysical masers: *Proceedings of the Conference*, Arlington, VA, USA, 1992, Springer-Verlag, pp. 195.
MacLeod, G.C. and Gaylard, M.J.: 1996, *MNRAS* **280**, 868.
Ouyed, R., Clarke, D.A. and Pudritz, R.E.: 2003, *ApJ* **582**, 292.

ULTRA-COMPACT H II REGIONS

M.G. HOARE
School of Physics and Astronomy, University of Leeds, UK; E-mail: mgh@ast.leeds.ac.uk

(Received 16 April 2004; accepted 15 June 2004)

Abstract. This article reviews how observations of dense molecular gas can help us understand the dynamics and evolution of ultra-compact H II (UCHII) regions. Particular attention is paid to the cometary objects, which it is argued are likely to be the dominant class. Thermal molecular lines in emission and absorption are considered as well as the mainly dynamical information from masers. Most of the UCHII regions studied at high resolution have hot cores ahead them which makes interpretation of molecular observations much more complex. Significant support for the champagne flow model for cometary regions can been seen, adding to the evidence from studies of the ionised gas. How the prevalence of cometary objects may fit into a triggered, sequential picture of massive star formation is outlined.

Keywords: star formation, H II regions, molecular clouds

1. Introduction

H II regions are the most obvious manifestation of massive star forming regions. Free–free emission from the ionised gas is strong at radio continuum wavelengths and, being unaffected by dust extinction, is easily seen right across our galaxy and in others. The simple overall picture is that H II regions begin in very dense gas and expand to larger sizes driven by over-pressure over time until the OB star joins the field population as the surrounding interstellar material is dispersed (Habing and Israel, 1979). Hence, for the purposes of studying massive star formation it is the youngest, so-called ultra compact H II regions (UCHIIs), and the dense molecular gas that surrounds them that attracts the most attention. These have $n_e > 10^4$ cm^{-3} and sizes less than 0.1 pc or <20 arcseconds at typical distances of >1 kpc (Churchwell, 1990).

The bulk of the luminosity of these embedded objects emerges in the far-IR and is a good measure of the total bolometric flux of the exciting stars. In addition, the radio continuum flux measures the total luminosity in the Lyman continuum. These two numbers, together with an assumption of the spectral energy distribution for the stars, can be used to determine their spectral type and mass. Such simple analytic tools have meant that H II regions are constantly used to study the massive star formation rate and efficiency in a wide range of environments.

To make use of the numbers of H II regions found in surveys we must first understand how they evolve and where they fit in the whole evolutionary scheme

Astrophysics and Space Science **295:** 203–215, 2004.

for young massive stars. By their very nature they form when the Lyman continuum radiation begins to emerge from the newly formed young star in sufficient quantities. Massive stars begin core hydrogen burning whilst still accreting (e.g. Yorke, 1986). There are many examples of luminous IR objects in massive star-forming regions that have not yet developed H II regions – the so-called massive young stellar objects (MYSOs). These invariably drive bipolar molecular outflows, which in most models of such flows requires active ongoing accretion. Compact H II regions, on the other hand, do not usually exhibit such phenomena and so one expects that accretion is over by this stage.

Some confusion can often occur between the UCHII region and the MYSO stages since the latter also exhibit radio continuum emission, but in this case it arises from an ionised stellar wind. The radio spectral index is usually close to the value of +0.6 expected for constant velocity, optically thick, winds (Wright and Barlow, 1975). However, compact H II regions can be optically thick up to quite high frequencies and variable density gradients in both types of objects can mean that their spectral indices overlap over a range of values from 0.6 to 2. At high resolution the MYSOs can be resolved into either radio jets (e.g. Rodríguez et al., 1994; Martí et al., 1998) aligned with larger scale outflows, or equatorial winds perpendicular to the bipolar flows (Hoare et al., 1994; Hoare, 2002). The latter may arise from material on the surfaces of accretion discs being radiatively driven away by the central star (Drew et al., 1998). To conclusively distinguish between a MYSO and a UCHII region it is necessary to examine the IR recombination line profiles. Those from UCHII regions have widths of a few tens of km s^{-1} arising from dynamical motions at around the sound speed of the photoionised gas, whilst those from MYSO winds are several hundreds of km s^{-1} broad reflecting the escape speed of their origin close to the central star (e.g. Bunn et al., 1995). Objects with broad radio recombination lines (De Pree et al., 2004; Sewilo et al., 2004) such as NGC 7538 IRS 1 may represent a transition stage as the initial ionisation of the surroundings results in a very compact, often bipolar radio source.

In the light of above, it is interesting to ask whether the start of the UCHII phase is directly linked to the end of the accretion phase. Walmsley (1995) has proposed that the high-infall rate onto a massive young star is sufficient to quench the formation of an H II region until the infall rate drops below a critical value (see also Keto, 2003). The ionised stellar wind can similarly prevent the emergence of Lyman continuum radiation since it is likely to recombine fairly close to the star (e.g. Höflich and Wehrse, 1987). These explanations rely on basically symmetric infall or outflow, which is almost certainly not the case. The ionized wind could be fed by material in the accretion disc (Drew et al., 1998). Any scenario that relies on the accretion rate would then result in the beginning of the UCHII region phase being coincident with a drop-off in the accretion rate.

Many questions also remain once the UCHII region begins to evolve. The dynamics of the region as it expands is the subject of much debate. This is particularly important since the main effect of H II regions is to first compress and then disperse the molecular cloud. The compression can trigger further molecular clumps to collapse leading to a wave of sequential star formation (Elmegreen and Lada, 1977), whilst ultimately the dispersal brings the end of the star formation episode (Franco et al., 1994). These competing effects due to H II region expansion will partly determine the overall star formation efficiency of a region. A major factor in how this expansion proceeds is the nature of the dense molecular gas that the H II region is expanding into: its overall density, density distribution and clumpiness. It is this context that will be explored in this review.

2. H II Region Dynamics

In the simplest picture H II regions expand into the surrounding neutral medium due to overpressure of the hot ionised gas at a rate depending on the density of the ambient medium. Spherical H II regions result from the assumption that the ambient medium is assumed to be uniform and smooth; neither of which is likely to be a good approximation in reality.

Departures from spherical symmetry were seen to be common when radio interferometers began to provide extinction-free views of H II regions. Israel (1978) found many objects with what he deemed a blister appearance. From a comparison of radio recombination line and CO velocities he also concluded that H II regions form near the edges of molecular clouds with ionised gas flowing away from the cloud. In a series of papers concluding with Yorke et al. (1983) this picture was explored hydrodynamically and termed a champagne flow as the ionised gas accelerated away from the cloud once the Strömgren sphere had reached the edge of a uniform density cloud.

In the meantime, higher resolution radio studies were uncovering large numbers of ultra-compact H II regions and many with what was now called a cometary morphology (Wood and Churchwell, 1989a (WC); Kurtz et al., 1994). Fich (1993) extended Israel's original work with a systematic VLA study of visible H II regions and again found that about a third showed such cometary morphology. In many cases, objects that were unresolved by Wood and Churchwell were shown to be cometary at higher resolution (e.g. M17 UC-1 Felli et al., 1984). Similarly, those broken up into irregular sources by the VLA at high resolution are often parts of larger cometary structures when seen at lower resolution. Kurtz et al. (1999) and Kim and Koo (2001) performed a low-resolution VLA study of many of the WC UCHII regions and found they were often associated with more extended ionised structures. They argued that this meant they were not, in fact, ultra-compact at all. However, the limited range of spatial scales in a given VLA configuration means it is often difficult to tell whether the ultra-compact and extended sources form a continuous structure or result from the superposition of different objects at different stages. It would not be surprising to find the latter situation in a sequential star formation picture. IR images, which are sensitive to all spatial scales down to their resolution, are often better for judging between these scenarios morphologically.

Wood and Churchwell (1989a,b) also began a debate about the lifetime of the ultra-compact phase arrived at from counting IRAS sources with the colours of known UCHII regions. They argued for a lifetime of 10^5 years rather than the 10^4 years predicted for simple Strömgren expansion into a medium with a density of 10^4 cm^{-3}. This approach was shown to greatly overestimate the number of UCHIIs by including both more evolved H II regions (Codella et al., 1994), clouds heated by lower luminosity B stars (Ramesh and Sridharan, 1997) and it will also include the MYSOs as well.

However, new dynamical models were developed to address both aspects of the cometary morphology and prolonged lifetime. The bow shock model was proposed in which the stellar wind from a moving OB star interacts with the molecular cloud (Mac-Low et al., 1991; Van Buren et al., 1990). The UCHII region is then set up in the approximately parabolic standing shock between the wind and the moving molecular cloud that produces a long-lived cometary structure. Turner and Matthews (1984) had previously appealed to stellar winds in a stationary situation to compress a spherical H II region and trap the ionisation front and prolong an ultra-compact phase.

Redman et al. (1998) invoked a mass-loading scenario giving rise to a recombination front rather than an ionization front. An asymmetric distribution of mass-loading centres then causes a cometary appearance in the same way as the champagne flow utilises an off-centre density gradient. De Pree et al. (1995) in the meantime had simply appealed to a greater density in the molecular environment to reduce the expansion rate. Increasing the ambient density up to 10^7 cm^{-3} significantly increased the lifetime of the simple Strömgren expansion. Xie et al. (1996) showed that this produced too large an emission measure in most cases and appealed to strong turbulent pressure instead to produce the confinement. This could now be seen to fit in with the turbulence-dominated mode of massive star formation proposed by McKee and Tan (2003).

We can distinguish between these various dynamical models for UCHII regions by examining the velocity structure within the ionized gas and how it relates to the surrounding molecular gas.

3. Velocity Structure in the Ionised Gas

The preponderance of cometary UCHIIs inspired many high-resolution radio recombination line studies of their velocity structure with very mixed conclusions. Wood and Churchwell (1991), Van Buren and Mac Low (1992) and Afflerbach et al. (1994) studied the archetypal cometary G29.96–0.02 and concluded that it was consistent with the bow shock model, although no detailed comparison with tailored models was attempted. Other radio recombination line studies found evidence for champagne flows in some cometary objects (e.g. Garay et al., 1994). W3(OH) – the other archetypal UCHII region – exhibits a well-defined velocity gradient interpreted as a champagne flow (Keto et al., 1995).

Cyganowski et al. (2003) argued that the two cometary regions in DR 21, whose velocities at their heads differ by 9 km s^{-1}, support the bow shock model. However, the velocity patterns along the axes clearly show that they are oppositely orientated. Hence, in a champagne interpretation the head of the southern source is expanding towards us whilst that of the northern source is expanding away from us. The head velocities sit either side of the ambient cloud velocity with the signs one would expect in this scenario. Cyganowski et al. did not consider that the ionisation front at the head of a champagne flow will still expand into the cloud. The usual physics applies here and this expansion at the head can be seen in the Bodenheimer et al. (1979) models. Therefore the expected velocity field for a champagne flow is more akin to the 'hybrid' model in Cyganowski et al., which the wider velocity field supports.

Radio recombination lines are relatively weak, which means there is a trade-off between spatial resolution and sensitivity, and the lower frequency lines are also strongly affected by pressure broadening. An alternative approach is to use the much stronger near-IR recombination lines, which are unaffected by pressure broadening. Lumsden and Hoare (1996, 1999) performed such a study, again on G29.96–0.02. Comparison of the data was made with approximations to bow-shock and champagne flow models, with much better agreement being found with the latter. The velocity structure at the head was very marked and not easily explainable by any one of the models without modification.

Better quality near-IR spectroscopy along the axis of G29.96-0.02 has recently been obtained by Martin-Hernández et al. (2003). This shows a sharp break in the velocity structure in front of the head of the cometary region, which may indicate that it is not part of the dynamical structure shaping the UCHII region itself. This is supported by the velocity structure seen in the other four cometary objects (G35.20–1.74, G43.89–0.78, G45.45+0.06, G19.61–0.23A) by Hoare et al. (2003). These had similar structure within the body of the nebula, but none had such extreme velocity and width gradients ahead of the apex.

More evolved sources have also had their velocity structure investigated, usually via velocity mapping of the Hα line (e.g. Miville-Deschenes et al., 1995; Priestley, 1999). The large cometary objects have velocity fields that also are consistent with a champagne flow. This makes the champagne model more attractive for the ultra-compact sources as well, since there is then no need to invoke a change in the physical processes controlling the dynamical evolution at some rather arbitrary point in the expansion.

Another direct way to probe the velocity structure of UCHII regions is to make proper motion measurements. Kawamura and Masson (1998) measured an expansion velocity of about 4 km s^{-1} in W3 (OH) and dynamical age of 2300 years. This is consistent with simple Strömgren expansion into neutral media with high, but not unreasonable, densities of about 10^7 cm^{-3}. It also predicts that this object would remain in the UCHII phase for about 10^5 years. Acord et al. (1998) found a very high expansion velocity of 35 km s^{-1} for G5.89, although the shell of this object is less well-defined and it has complicated outflow signatures as well. A key to confirming both the Strömgren expansion of the ionisation fronts and champagne flows down density gradients is the density and distribution of the material in the surrounding molecular cloud.

4. Dense Molecular Gas

4.1. Thermal Emission Lines

There have been numerous studies of molecular line emission from the environs of UCHII regions, mostly with single-dish observations. In the light of the VLA continuum surveys there were studies of UCHII samples in NH_3 by Churchwell et al. (1990) and $C^{34}S$ by Cesaroni et al. (1991), CH_3C_2H by Fontani et al. (2002) and more recently in $C^{17}O$ by Hofner et al. (2000). Dense (10^{5-6} cm^{-3}) molecular gas was associated with all the ultra-compact objects as expected for regions where massive stars have recently formed.

Maps in ^{13}CO and various CS transitions of several complexes containing UCHII regions have been made by Kim and Koo (2003). These show that for some of the cometary objects the densest part of the molecular cloud is indeed at the head of the cometary region seen in radio continuum as expected for the champagne flow, e.g. see G12.21–0.10 (Figure 1). However, the typical spatial resolution of single-dish observations of about 30–60 arcseconds is usually insufficient and the dense gas appears to peak on the UCHII region itself. Kim and Koo did find that the density and velocity structure of the diffuse extended ionised and molecular gas was consistent with a champagne interpretation.

Equivalent information on the density distribution at somewhat better spatial resolution is available from submillimetre dust continuum emission maps (Mueller et al., 2002; Thompson et al., 2002). These again show very centrally condensed clouds centred on the UCHII regions themselves, but if there is an offset of the peak density then it is usually ahead of the cometary apex (again see G12.21–0.10). Hatchell and van der Tak (2003) found that the average radial density distribution derived from resolved dust cores containing UCHII regions was consistent with the $r^{-1.5}$ expected for free-fall collapse.

In about half of all UCHII regions it is found that they are associated with warm molecular gas as evidenced by highly excited NH_3 (Cesaroni et al., 1992), CS (Olmi and Cesaroni 1999), methanol masers (Walsh et al. 1998) or other molecular tracers of the earlier hot core phase (Hatchell et al., 1998). Higher-resolution data showed that these hot cores were usually spatially offset from the UCHII regions by a few arcseconds (Cesaroni et al., 1994). Where these were associated with

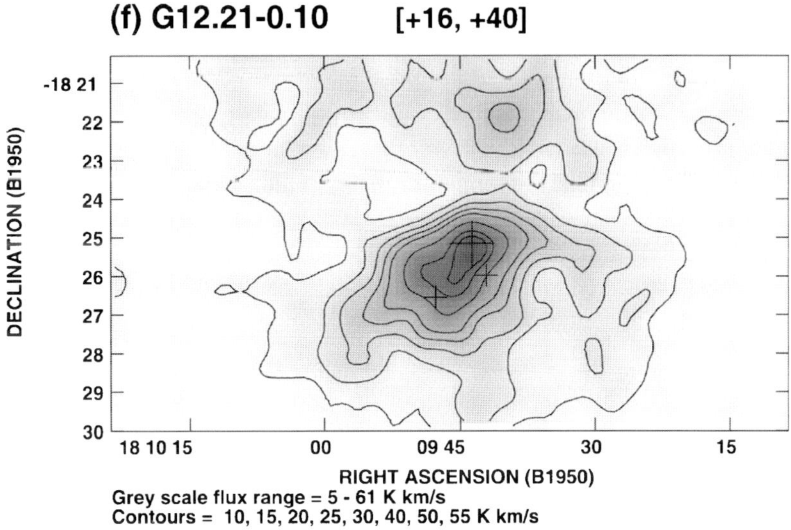

Figure 1. ^{13}CO integrated emission towards the UCHII region G12.21–0.01 from Kim and Koo (2003). The UCHII region (large cross) is cometary with its tail pointing south down the density gradient seen in the molecular gas consistent with a champagne flow. More evolved H II regions (smaller crosses) are located in the less dense gas consistent with a wave of sequential star formation moving northward into denser gas.

cometary regions then the hot cores were always located ahead of the apex of the region. Many cometary regions have been found to show signs of earlier stages of massive star formation such as maser sources or massive YSOs as well as hot cores in the dense regions a few arcseconds ahead of the cometary region (e.g. Hofner et al., 1994; Hofner and Churchwell, 1996). This is to be expected in a sequential champagne picture where expansion of the region into the dense gas will trigger further massive star formation.

Although this provides support for the champagne model it has hampered direct testing since most subsequent interferometric studies of the dense molecular gas around UCHII regions has since been focused on those containing the hot cores. These very young sources are of great interest, but the gravitational collapse that has led to their formation can significantly alter the density and velocity distribution in the immediate vicinity of the apex of the cometary UCHII regions.

A good example of this is again G29.96–0.02. Interferometric observations of ^{13}CO 1–0 and $C^{18}O$ 1-0 by Pratap et al. (1999) and Olmi et al. (2003) reveal gas densities of about 5×10^5 cm^{-3}. The density does peak up at, or just ahead of, the head of the cometary H II region, but this is also the location of the hot core. Interestingly, blue-shifted CO and HCO+ 1-0 emission (Maxia et al., 2001) is seen in the tail region, which would be consistent with molecular gas being entrained within the champagne flow seen in the ionised gas. However, Maxia et al. also see HCO+ 1-0 in absorption at a velocity that is redshifted with respect to main molecular cloud and hence appears to be infalling towards the UCHII region. They interpret this as a manifestation of infall onto the hot core rather than the H II region itself.

There are many other aspects that could complicate interpretation of molecular signatures around UCHII regions. A compressed neutral shell should be present between the shock ahead of the ionisation front, which will generate strong molecular line emission as part of a photon-dominated region. Chemical abundances are also drastically altered in the vicinity of the hot core. Hence, any interpretation has to be carefully considered and detailed modelling needed to inform this.

High resolution observations of G34.24 by Watt and Mundy (1999) show $C^{18}O$ 1-0 tracing an arc around the head of this cometary region where one would expect the dense gas in a champagne interpretation. The derived densities are very high (10^7 cm^{-3}) several arcseconds from the UCHII region. In this object the traditional hot core tracer of methyl cyanide peaks up right on the apex of the cometary region. Watt and Mundy interpret this emission as arising from dense molecular gas externally heated by the UCHII region rather than internally heated by a new young massive star.

W3 (OH) has been observed at high resolution by Wyrowski et al. (1999) principally to study the peculiar source W3 (H_2O) in the vicinity. Several dense gas tracers such as H_2CO and CH_3OH show an arc of emission wrapped around the western end of the UCHII region. Such a pattern is consistent with a champagne flow picture for this source, which is discussed further in the next section.

4.2. MOLECULAR ABSORPTION LINES

A useful alternative probe of dense molecular gas around UCHII regions is to use lines in absorption against the strong continuum background. This gives additional constraints due to the particular geometry necessary to generate them. The H_2CO lines at 2 and 6 cm have been observed at high resolution in absorption against W3(OH) and W58 C1 (Dickel and Goss, 1987; Dickel et al., 2001). Both of these objects have a cometary structure and the H_2CO line optical depth maps both show peaks consistent with the densest gas being around the head of the cometary region (Figure 2). Densities of 6×10^4 cm^{-3} were derived from the strengths of the absorption lines. The authors themselves interpret other details in the line structure as arising from possible outflows in these regions.

Ammonia lines in both emission and absorption have also been seen towards many UCHII regions. Keto (2002), in one of his many detailed studies of G10.6–0.4, mapped ammonia absorption at high resolution and compared the velocity structure to that seen in the ionised gas via radio recombination lines. He interpreted the fact that both the molecular and ionised lines have larger redshifts towards the centre of this object than the edges as being due to approximately radial infall – a claim that is also made for a couple of other compact symmetric objects.

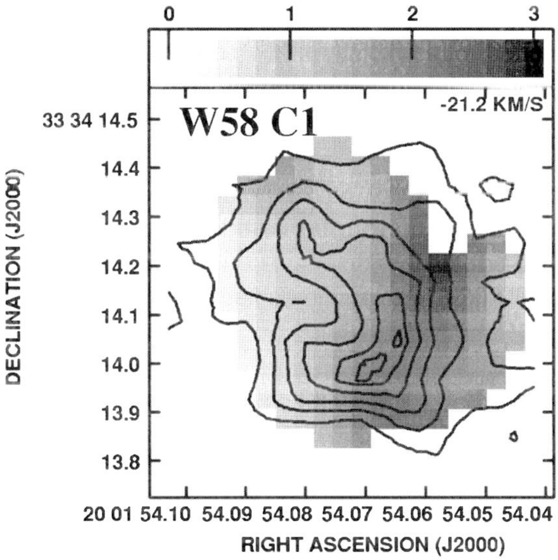

Figure 2. Image of the line centre optical depth for the 2 cm H_2CO absorption line against the continuum (shown in contours) for W58 C1. Note the optical depth peaks around the western edge where one would expect the densest gas in a champagne flow picture for this cometary object. From Dickel et al. (2001).

4.3. Maser lines

Maser lines are, by their very nature, difficult to use as a diagnostic of the physical conditions of the dense molecular gas around UCHII regions. These objects are one of the best hunting grounds for new maser transitions, which may help understand the various maser pumping mechanisms. OH is the most common masing species associated with UCHII regions and excited OH transitions at 6.0 and 13.4 GHz have been widely seen (e.g. Baudry and Desmurs, 2002). Cragg et al. (2002) have used the fact that CH_3OH and OH masers sometimes appear to originate in the same gas to constrain the conditions needed to excite both. These require cool ($T < 100$ K), dense (10^{5-8} cm^{-3}) gas to be irradiated by warm ($T > 100$ K) dust emission; conditions which are consistent with what we know about UCHII regions.

More importantly, the strong compact maser spots can be used as a dynamical probe, since their 3D motion can be traced via radial velocities and tangential proper motions with VLBI. The most well-studied object is again W3(OH) with proper motion speeds of several km s^{-1} being commonly found. Bloemhof et al. (1992) concluded that the OH proper motions were consistent with an expansion of about 3 km s^{-1}, in excellent agreement with the radio continuum measurements of Kawamura and Masson (1998). However, the maser proper motions were only measured in a relative sense and with a reasonable choice of reference motion the pattern can be made to flow generally in the NE direction towards the low density tail. Bloemhof et al. concluded that this, and a similar pattern derived by a diagonalisation method (Bloemhof, 1993), resembled a bow-shock type velocity field as predicted by Van Buren et al. (1990). The latter is relative to the central star and not the molecular cloud, whilst a similar velocity field is also predicted by the champagne flow in the molecular cloud (and central star) reference frame.

Moscadelli et al. (2002) found similar amplitude motions in CH_3OH masers, but interpreted many of the motions as arising in bipolar outflows from embedded sources around the edge of the UCHII region rather than due to the dynamics of the H II region itself. The long running debate as to whether CH_3OH masers arise directly as a result of the UCHII region or are formed in the immediate environs of other nearby massive YSOs and/or hot cores has not yet been fully settled (e.g. Beuther et al., 2002). Systematic, detailed, high-resolution studies at a variety of wavelengths are needed to unravel the dense clustering of the various stages of the early evolution of massive stars.

5. Conclusions

The majority of the evidence now indicates that the dynamics, morphology and lifetimes of UCHII regions can be explained within the broad terms of the champagne flow picture. Dense and/or turbulent molecular gas at the birth sites of massive stars may prolong the lifetimes over and above the 10^4 years that used to be typically assumed, but not to the extent that Wood and Churchwell (1989b) originally supposed. There appears to be no need to invoke the large proper motions that underly the bow shock scenario to explain the large fraction of cometary objects. However, the effect of stellar winds on the dynamics and morphology does need to be included and probably explains the limb-brightened nature of many UCHII regions.

The final proof will come from a detailed comparison of hydrodynamic champagne flow models including stellar winds and realistic density distributions with all the available data for a number of objects (Arthur and Hoare, in preperation). Large systematic samples of UCHII regions will also answer the lifetime question once and for all. A survey of red MSX sources currently underway (Lumsden et al., 2003; Hoare et al., 2004) should be close to complete for all the UCHII regions in the galaxy. Significant progress is being made in the study of the exciting stars themselves through high-resolution IR imaging and spectroscopy (Watson and Hanson, 1997; Martin-Hernández et al., 2003). Hence, it will soon be possible to include realistic stellar parameters into models of UCHII regions as well.

Molecular gas with densities of 10^{4-7} cm^{-3} has been observed around UCHII regions for some time. However, the higher resolution needed to really help test the dynamical evolution of UCHII regions and their role in triggering and dispersing the molecular cloud is only just beginning. There is an urgent need to observe UCHII regions, especially cometary ones, that do not have hot cores and/or MYSOs in close proximity. This would allow a much cleaner test of the dynamical models in a molecular setting before further gravitational collapse has been set in motion and drastically altered the picture. A high-dynamical range and number of tracers is also needed to trace the molecular environment from the dense head region to the diffuse tail. Such demanding observations will require ALMA.

Maser observations have the potential to deliver three-dimensional tests of the dynamics in the dense gas in the immediate shell around UCHIIs. However, most studies so far have only observed relative proper motions. With phase referencing now available for VLBI, future observations should be made in this mode so that absolute proper motions can be derived and the vast uncertainty in the interpretation of maser motions removed.

If the dominance of cometary objects flowing down density gradients is confirmed as the norm for UCHII regions then it begs the question as to why massive stars rarely form right at the densest peaks in the very centres of molecular clouds. This could be explained if triggering was necessary to initiate collapse in most cases. A pressure wave from whatever source; spiral density wave, supernova remnant or previous generation of massive star formation, will necessarily impact on a cloud from a particular direction. As the wave moves up the pre-existing density gradient at some point the compression will be enough to trigger gravitational collapse and massive stars will rapidly form. Although the collapse will concentrate the density distribution locally, when the UCHII region phase begins the expanding ionised zone will quickly enter the overall density gradient and flow predominately down it. The original triggering wave may have moved on into the cloud creating further star formation or it may have stalled. In the latter case, sequential triggering will then take over and the expansion at the head of the cometary UCHII region will carry on the task. Perhaps only in a spontaneous collapse picture would one expect massive stars to mostly form at the densest peaks of a molecular cloud and that the subsequent UCHII regions would then be more symmetrical on the whole. The triggered cometary picture outlined above needs to be explored by more global hydrodynamical models.

References

Acord, J.M., Churchwell, E. and Wood, D.O.S.: 1998, *ApJ* **495**, L107.
Afflerbach, A., Churchwell, E., Hofner, P. and Kurtz, S.: 1994, *ApJ* **437**, 697.
Baudry, A. and Desmurs, J.F.: 2002, *A&A* **394**, 107.
Beuther, H., et al.: 2002, *ApJ* **390**, 289.
Bloemhof, E.E., Reid, M.J. and Moran, J.M.: 1992, *ApJ* **397**, 500.
Bloemhof, E.E.: 1993, *ApJ* **406**, L75.
Bodenheimer, P., Tenorio-Tagle, G. and Yorke, H.W.: 1979, *ApJ* **233**, 85.
Bunn, J.C., Hoare, M.G. and Drew, J.E.: 1995, *MNRAS* **272**, 346.
Cesaroni, R., Walmsley, C.M., Kömpe, C. and Churchwell, E.: 1991, *A&A* **252**, 278.
Cesaroni, R., Walmsley, C.M. and Churchwell, E.: 1992, *A&A* **256**, 618.
Cesaroni, R., Churchwell, E., Hofner, P., Walmsley, C.M. and Kurtz, S.: 1994, *A&A* **288**, 903.
Churchwell, E., Walmsley, C.M. and Cesaroni, R.: 1990, *A&AS* **83**, 119.

Churchwell, E.: 1990, *A&A Rev* **2**, 79.
Codella, C., Felli, M. and Natale, V.: 1994, **284**, 233.
Comeron, F.: 1997, *A&A* **326**, 1195.
Cragg, D.M., Sobolev, A.M. and Godfrey, P.D.: 2002, *MNRAS* **331**, 521.
Cyganowski, C.J., Reid, M.J., Fish, V.L. and Ho, P.T.P.: 2003, *ApJ* **596**, 344.
De Pree, C.G., Wilner, D.J., Mercer, A.J., Davis, L.E., Goss, W.M. and Kurtz, S.: 2004, *ApJ* **600**, 286.
De Pree, C.G., Rodriguez, L.F. and Goss, W.M.: 1995, *Rev. Mex. Astron. Astrofis.* **31**, 39.
Dickel, H.R. and Goss, W.M.: 1987, *A&A* **185**, 271.
Dickel, H.R., Goss, W.M. and De Pree, C.G.: 2001, *AJ* **121**, 391.
Drew, J.E., Proga, D. and Stone, J.M.: 1998, *MNRAS* **296**, L6.
Elmegreen, B.G. and Lada, C.: 1977, *ApJ* **214**, 725.
Felli, M., Massi, M. and Churchwell, E.: 1984, *A&A* **136**, 53.
Fich, M.: 1993, *ApJS* **86**, 475.
Fontani, F., Cesaroni, R., Caselli, P. and Olmi, L.: 2002, *A&A* **389**, 603.
Franco, J., Shore, S.N. and Tenorio-Tagle, G.: 1994, *ApJ* **436**, 795.
Garay, G., Lizano, S. and Gomez, Y.: 1994, *ApJ* **429**, 268.
Habing, H.J. and Israel, F.P.: 1979, *ARAA* **17**, 345.
Hatchell, J., Thompson, M.A., Millar, T.J. and MacDonald, G.H.: 1998, *A&AS* **133**, 29.
Hatchell, J. and van der Tak, F.F.S.: 2003, *A&A* **409**, 589.
Hoare, M.G., Drew, J.E., Muxlow, T.B. and Davis, R.J.: 1994, *ApJ* **421**, L51.
Hoare, M.G.: 2002, in: P.A. Crowther (ed.), *Hot Star Workshop III: The Earliest Stages of Massive Star Birth*, ASP Conference Series, Vol. 267, p. 137.
Hoare, M.G., Lumsden, S.L., Busfield, A.L. and Buckley, P.: 2003, in: S.J. Arthur and W.J. Henney (eds.), *Rev. Mex. Astron. Astrofis. Ser. Conf. 15*, Winds, Bubbles and Explosions, p. 172.
Hoare, M.G., Lumsden, S.L., Oudmaijer, R.D., Busfield, A.L. and King, T.L.: 2004, in: D. Clemens, T. Brainerd and R. Shah (eds.), *Milky Way Surveys: The Structure and Evolution of Our Galaxy*, ASP Conference Series in press.
Höflich, P. and Wehrse, R.: 1987, *A&A* **185**, 107.
Hofner, P., Wyrowski, F., Walmsley, C.M. and Churchwell, E.: 2000, *ApJ* **536**, 393.
Hofner, P. and Churchwell, E.: 1996, *A&AS* **120**, 283.
Hofner, P., Kurtz, S., Churchwell, E., Walmsley, C.M. and Cesaroni, R.: 1994, *ApJ* **429**, L85.
Israel, F.P.: 1978, *A&A* **90**, 769.
Kawamura, J.H. and Masson, C.R.: 1998, *ApJ* **509**, 270.
Keto, E.R.: 2002, *ApJ* **568**, 754.
Keto, E.R.: 2003, *ApJ* **599**, 1196.
Keto, E.R., Welch, W.J., Reid, M.J. and Hp, P.T.P.: 1995, *ApJ* **444**, 765.
Kim, K.-T. and Koo, B.-C.: 2001, *ApJ* **549**, 979.
Kim, K.-T. and Koo, B.-C.: 2003, *ApJ* **596**, 362.
Kippenhahn, R. and Meyer-Hofmeister, E.: 1977, *A&A* **54**, 539.
Kurtz, S., Churchwell, E. and Wood, D.O.S.: 1994, *ApJS* **91**, 659.
Kurtz, S., Watson, A.M., Hofner, P. and Otte, B.: 1999, *ApJ* **514**, 232.
Lefloch, B. and Lazereff, B.: 1994, *A&A* **289**, 559.
Kim, K.-T. and Koo, B.-C.: 2003, *ApJ* **596** 362.
Lumsden, S.L. and Hoare, M.G.: 1996, *ApJ* **464**, 272.
Lumsden, S.L. and Hoare, M.G.: 1999, *MNRAS* **305**, 701.
Lumsden, S.L., Hoare, M.G., Oudmaijer, R.D. and Richards, D.: 2002, *MNRAS* **336**, 621.
Mac Low, M.-M., Van Buren, D., Wood, D.O.S. and Churchwell, E.: 1991, **369**, 395.
Martí, J., Rodríguez, L.F. and Reipurth, B.: 1998, *ApJ* **502**, 337.
Martin-Hernández, N.L., Bik, A., Kaper, L., Tielens, A.G.G.M. and Hanson, M.M.: 2003, *A&A* **405**, 175.

Maxia, C., Testi, L., Cesaroni, R. and Walmsley, C.M.: 2001, *A&A* **371**, 286.
McKee, C.F. and Tan, J.C.: 2003, *ApJ* **585**, 850.
Miville-Deschenes, M., Joncas, G. and Durand, D.: 1995, *ApJ* **454**, 316.
Moscadelli, L., Menten, K.M., Walmsley, C.M. and Reid, M.J., 2002, *ApJ* **564**, 813.
Mueller, K.E., Shirley, Y.L., Evans, N.J. II, and Jacobson, H.R.: 2002, *ApJS* **143**, 469.
Olmi, L., Cesaroni, R., Hofner, P., Kurtz, S., Churchwell, E. and Walmsley, C.M.: 2003, *A&A* **407**, 225.
Olmi, L. and Cesaroni, R.: 1999, *A&A* **352**, 266.
Pratap, P., Megeath, S.T. and Bergin, E.A.: 1999, *ApJ* **517**, 799.
Priestley, C.: 1999, PhD Thesis, University of Leeds.
Ramesh, B. and Sridharan, T.K.: 1997, *MNRAS* **284**, 1001.
Redman, M.P., Williams, R.J. and Dyson, J.E.: 1998, *MNRAS*, **298**, 33.
Rodríguez, L.F., Garay, G., Curiel, S., Ramírez, S., Torrelles, J.M., Gómez, Y. and Velázquez: 1994, *ApJ* **430** L65.
Sewilo, M., Churchwell, E., Kurtz, S., Goss, W.M. and Hofner, P.: 2004, *ApJ* **605**, 285.
Thompson, M.A., Hatchell, J., MacDonald, G.H. and Millar, T.J.: 2002, in: P. A. Crowther (ed.), *Hot Star Workshop III: The Earliest Stages of Massive Star Birth*, ASP Conference Series, Vol. 267, p. 429.
Turner, B.E. and Matthews, H.E.: 1984, *ApJ* **277**, 164.
Van Buren, D., Mac Low, M.-M., Wood, D.O.S. and Churchwell, E.: 1990, *ApJ* **353**, 570.
Van Buren, D. and Mac Low, M.-M.: 1992, *ApJ*, **394**, 534.
Xie, T., Mundy, L.G., Vogel, S.N. and Hofner, P.: 1996, *ApJ* **473**, L131.
Walmsley, M.: 1995, in: S. Lizano and J. Torrelles (eds.), *Rev. Mex. Astron. Astrofis. Ser. Conf. 1, Circumstellar Disks, Outflows, and Star Formation*, DF: Inst. Astron., UNAM, Mexico, p. 137.
Walsh, A.J., Burton, M.G., Hyland, A.R. and Robinsonm, G.: 1998, *MNRAS* **301**, 640.
Watson, A.M. and Hanson, M.M.: 1997, *ApJ* **490**, L195.
Watt, S. and Mundy, L.G.: 1999, *ApJS* **125**, 143.
Wood, D.O.S. and Churchwell, E.: 1989a, *ApJS* **69** 831.
Wood, D.O.S. and Churchwell, E.: 1989b, *ApJ* **340**, 265.
Wood, D.O.S. and Churchwell, E.: 1991, *ApJ* **372**, 199
Wright, A.E. and Barlow, M.J.: 1975, *MNRAS* **170**, 41.
Wyrowski, F., Schilke, P., Walmsley, C.M. and Menten, K.M.: 1999, *ApJ* **514**, L43.
Yorke, H.W., Tenorio-Tagle, G. and Bodenheimer, P.: 1983, *A&A*, **127**, 313.
Yorke, H.W.: 1986, *ARAA* **24**, 49.

MASER ACTION IN METHANOL TRANSITIONS

MAXIM VORONKOV[1], ANDREJ SOBOLEV[2], SIMON ELLINGSEN[3], ANDREI OSTROVSKII[4] and ALEXEI ALAKOZ[5]

[1]*Australia Telescope National Facility, Australia, Astro Space Centre, Russia;*
E-mail: Maxim.Voronkov@csiro.au
[2]*Ural State University, Russia*
[3]*University of Tasmania, Australia*
[4]*Ural State University, Russia*
[5]*Astro Space Centre, Russia*

(Received 16 April 2004; accepted 15 June 2004)

Abstract. We report the detection with the ATCA of 6.7 GHz methanol emission towards OMC-1. The source has a size between 40″ and 90″, is located to the south-east of Ori-KL and may coincide in position with the 25 GHz masers. The source may be an example of an interesting case recently predicted in theory where the transitions of traditionally different methanol maser classes show maser activity simultaneously. In addition, results of recent search for methanol masers from the 25 and 104.3 GHz transitions are reported.

Keywords: masers – ISM: molecules

1. Introduction

Interstellar methanol masers are associated with regions of active star formation and masing transitions are traditionally divided into two classes Menten et al. (1991). Although masers of both classes often co-exist in the same star-forming region, Class I masers are normally seen apart from the strong continuum sources, while Class II masers are found close to them. The $5_1 - 6_0$ A$^+$ transition at 6.7 GHz (a Class II transition) produces the brightest known methanol maser (up to 5000 Jy). Maser emission from this transition is widespread and to date has been detected towards more than 400 sources. Among the Class I methanol masers there is the J_2–J_1 E series near 25 GHz. The strongest $5_2 - 5_1$ E 25 GHz maser (about 150 Jy) is observed towards the Orion molecular cloud (OMC-1) which is the best studied Class I source. The 25 GHz transitions are seen in absorption towards the archetypal Class II source W3(OH) (Menten et al., 1986) and the general assumption has been that other Class II sources will exhibit similar behaviour. To date only a few 25 GHz methanol masers have been detected and bright 25 GHz masers are thought to be rare (e.g., Menten et al., 1986). However, the survey of Menten et al. (1986) did not cover many sites which subsequent searches in other transitions have been found to be prominent Class I methanol maser sources. In this paper we report results of a pilot survey of 25 GHz masers targeted at Class I masers at 44 GHz with sufficiently well-known positions.

The difference between the two traditional classes is due to a difference in the pumping mechanism: strong Class I masers appear when the excitation processes are dominated by collisions while the strong Class II masers appear when the radiative excitation prevails (e.g., Cragg et al., 1992). Therefore, the masers in transitions of different classes are usually seen apart from each other. The work of Cragg et al. (1992) was limited to the case where conditions in the masing region are such that the radiative transitions between different torsionally excited states are not influential.

Further studies of methanol maser excitation have shown that the strong Class II masers are produced when excitation from the ground to the second torsionally excited state comes into effect (see, e.g., Sobolev and Deguchi, 1994; Sobolev et al., 1997a). Calculations show that the involvement of torsional transitions does not allow masers of different traditional classes to become strong simultaneously. However, the coexistence of inversion in some Class I and II transitions is predicted in some situations (e.g., the Class I $11_2 - 11_1$ E transition is present in the list of Class II methanol maser candidates of Sobolev et al., 1997b).

The general tendencies of methanol maser pumping are demonstrated in Figure 1 showing the dependence of brightness temperature in different masing (i.e., inverted) transitions on the temperature. In this model the collisional excitation is provided by the hydrogen molecules and the pumping radiation is produced by the dust intermixed with the gas within the cloud. The cloud has the following parameters: hydrogen number density 10^5 cm^{-3}, methanol abundance 10^{-5}, beaming factor (ε^{-1}) 10, specific (divided by linewidth) column density $10^{11.5}$ cm^{-3}s, temperature of the dust and gas are equal. Increasing the temperature in such a cloud changes the balance of excitation processes in favour of radiative ones. So, one should expect that the Class I masers should appear at low temperatures while at high temperatures the Class II masers should shine brightly. The figure shows

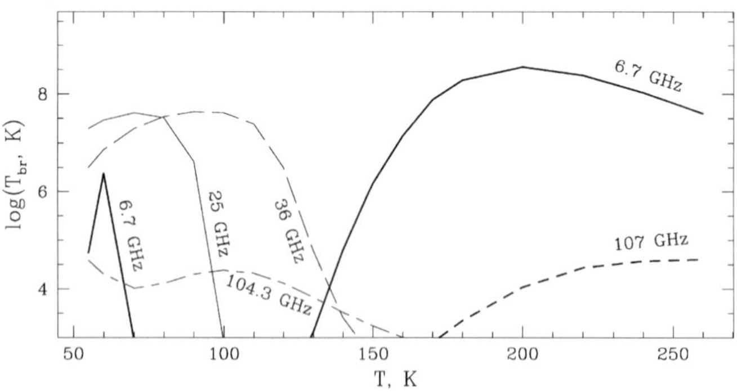

Figure 1. The model dependence of the brightness temperature of various methanol maser transitions on the temperature ($T = T_{\rm gas} = T_{\rm dust}$).

that this is correct, at least for well-known Class I masers at 25 GHz and 36 GHz and Class II maser at 107 GHz. In addition, the figure shows that there is some transition region when the strong Class I $5_2 - 5_1$ E maser can coexist with the weaker maser in Class II $5_1 - 6_0$ A^+ transition. This is not expected in the traditional classification scheme and we have attempted to test this prediction through observation. Further, Figure 1 shows that the maser at 104.3 GHz, which was predicted in theory (Voronkov, 1999), could be a representative of intermediate class. If the maser appears mostly under intermediate conditions then the detection rate in surveys targeted towards either Class I or Class II sources may be low.

2. Observational Results and Discussion

2.1. Search for 6.7 GHz Emission from the 25 GHz Maser Site in OMC-1

The Australia Telescope Compact Array (ATCA) was used to search for 6.7 GHz emission in OMC-1. Observations were made with the array in the compact H75 configuration on 10 June 2003. A weak spectral feature (about 0.2 Jy) has been detected (Figure 2, left). Using the newly installed 12-mm system at the ATCA, the 25 GHz maser was observed quasi-simultaneously with the 6.7 GHz observations. For both frequencies the correlator was configured to split the 4 MHz bandwidth into 1024 spectral channels. The spectral resolution at 25 GHz was 0.047 km s^{-1}. At 6.7 GHz each eight adjacent channels were averaged together reducing the effective spectral resolution to 1.4 km s^{-1}. Because the spatial resolution of these

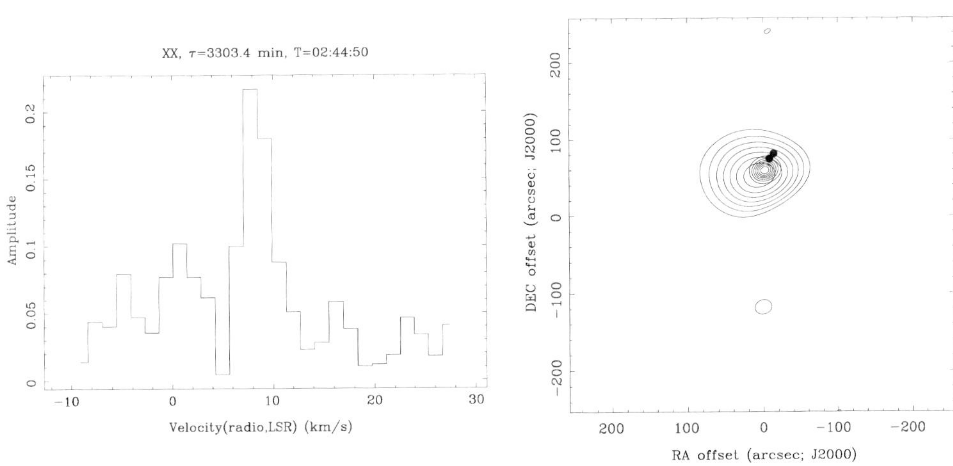

Figure 2. Left. The spectrum of detected 6.7 GHz emission in OMC-1. Amplitude is in Jy. Right. The combined map at 6.7 GHz and 25 GHz (a peak spectral channel corresponding to 8 km s^{-1} was used for imaging). Two filled circles represent the positions of BN and IRc2 objects.

get high-precision absolute and relative astrometry of the maser emission, we used the EVN in phase-referenced mode to make a large field of view observation of the region. The observations were made on 7 February 2002, using 4 EVN antennas which can observe at 6.7 GHz. The data were correlated with the EVN MarkIV data processor at JIVE. Because the maser emission is spread over a region larger than the primary beam of the Effelsberg telescope, the observations had to be made using two separate pointings.

3. Results and Discussion

Figure 1 shows the measured position of the methanol masers overlayed on an 4.8-GHz VLA continuum image of the region. Although a few sources are associated with known centres of activity, the majority of the masers do not show any obviously compact radio continuum. However, the presence of the methanol masers means that these are the locations of some sort of activity, presumably massive star formation. Interestingly, 800 μm observations made by Ladd et al. (1993) using the JCMT, show a compact continuum source at the same position as the masers in the southeast. The whole region has been surveyed for main line OH maser emission using the VLA (Argon et al., 2000,) and most of sites of methanol maser emission do not show any OH maser emission either. VLA observations of H_2O masers and thermal NH_3 do not detect any emission towards most of the methanol masers, but given the relatively small primary beam at 22 GHz, this is probably not

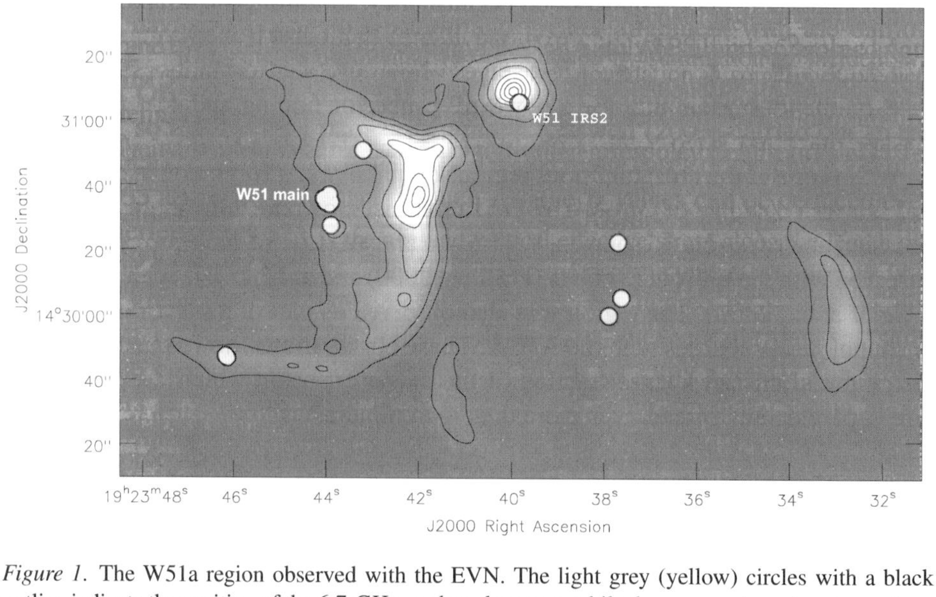

Figure 1. The W51a region observed with the EVN. The light grey (yellow) circles with a black outline indicate the position of the 6.7-GHz methanol masers, while the grey scale and contours show the 4.8 GHz continuum observed with the VLA (Mehringer, 1994).

surprising without targeted observations. The three sites of methanol emission in the southwest are located at the edge of a dark cloud seen in the MSX (4–26μm) (Egan and Price, 1996). This would indicate that these masers are associated with a very embedded region of star formation.

We suggest that the newly detected methanol masers represent either very young or very embedded (possibly both) massive stars or proto-stars. Follow-up observations using instruments such as BIMA and the VLA at 20 GHz will be needed to understand the origin of these sources.

Two of the sites of methanol maser emission in this region are well-known sources and have been studied extensively at many wavelength; W51 Main and W51 IRS2.

3.1. W51 MAIN

Figure 2 shows a close up of the W51 Main region, with the methanol positions from the current data overlayed on 8.4 GHz continuum emission as well as OH and

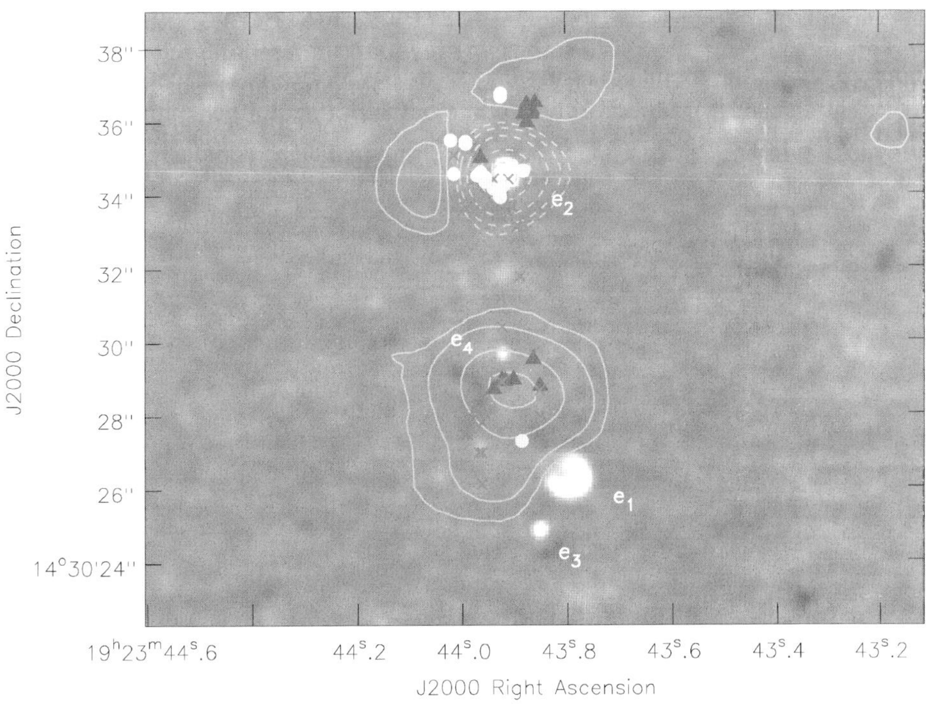

Figure 2. Detail of the W51 Main region. The light grey (yellow) circles indicated the position of 6.7-GHz methanol masers, the position of OH and water masers are indicated by crosses (red) and triangles (blue) (Argon et al., 2000; Imai et al., 2002). The grey scale shows the 8.4 GHz continuum emission observed using the VLA A array (Gaume et al., 1993), while the contours (green) show the integrated NH_3 (1,1) emission also observed with the VLA (Zhang and Ho, 1997).

NGC7538 IRS1 N: MODELING A CIRCUMSTELLAR MASER DISK

M. PESTALOZZI[1], M. ELITZUR[2], J. CONWAY[1] and R. BOOTH[1]
[1]*Onsala Space Observatory, Onsala, Sweden; E-mail: michele@oso.chalmers.se*
[2]*Department of Physics and Astronomy, University of Kentucky, Lexington, KY, USA*

(Received 16 April 2004; accepted 15 June 2004)

Abstract. We present an edge-on Keplerian disk model to explain the main component of the 12.2 and 6.7 GHz methanol maser emission detected toward NGC7538-IRS1 N . The brightness distribution and spectrum of the line of bright masers are successfully modeled with high amplification of background radio continuum emission along velocity coherent paths through a maser disk. The bend seen in the position–velocity diagram is a characteristic signature of differentially rotating disks. For a central mass of $30M_{\odot}$, suggested by other observations, our model fixes the masing disk to have inner and outer radii of ~270 and ~750 AU.

Keywords: Masers: Star formation

1. Introduction: Disks in High-Mass Star Formation Regions

Disks are expected to form during protostellar collapse, and low-mass stars seem to provide good observational evidence for the existence of disks (e.g., Qi et al., 2003; Fuente et al., 2003). The situation is less clear for high-mass stars. While in several cases velocity gradients in massive star forming regions have been detected on large scales (>10,000 AU, e.g. Sandel et al., 2003), evidence for compact disks on scales ≤1000 AU remains sparse. One possible example is IRAS20126 + 4104, studied by Cesaroni et al. (1999).

Class II methanol maser emission, a signpost of high-mass star formation (Minier et al., 2001), offers a potential indicator of disks since it often shows linear structures both in space and position–velocity diagrams (Norris et al., 1998; Minier et al., 2000). One of the most striking examples of a maser line in both space and velocity is found in NGC7538-IRS1 N (Minier et al., 1998). We present here the first quantitative Keplerian disk analysis of this maser without invoking the assumption of a single radius. We find compelling evidence for the disk interpretation in this case.

2. Basic Theory of the Model and Fit to the Data

Because of the remarkable agreement of the positions of the maser emissions at 6.7 and 12.2 GHz (2–3 mas), we draw the conclusion that the maser is the result of the amplification of a background continuum. Consider than an edge-on rotating disk at a distance D from the Sun and a point at radius $\rho = r/D$ along a path with displacement θ. The rotation velocity $V(\rho)$ and its line-of-sight component v obey $v/\theta = V/\rho = \Omega$, the angular velocity. In Keplerian rotation, $\Omega \propto \rho^{-3/2}$ with $\Omega_o = \Omega(\rho_o) = D\sqrt{GM/R_o^3}$, where $\rho_o = R_o/D$ is the outer radius and M the central mass. Assume a Gaussian frequency profile for the maser absorption coefficient and denote the radial variation of its magnitude at line center by the normalized profile $\eta(\rho)$ ($\int \eta\, d\rho = 1$). Then

$$\tau(\theta, v) = \tau_0 \int \eta(\rho) \exp\left[-\frac{1}{2}\left(\frac{v - \Omega(\rho)\theta}{\Delta v_D}\right)^2\right] \frac{d\rho}{\beta} \tag{1}$$

where $\beta = \sqrt{1 - (\theta/\rho)^2}$ and $d\rho/\beta$ is distance along the path. The width Δv_D is taken as constant and we set it to be the spectral line width of the maser emission at 12.2 GHz at $\theta = 0$. In our case, $\Delta v_D = 0.4$ km s^{-1}. This value could vary with ρ because of temperature variation and maser saturation. Saturation typically requires $\tau \geq 10$–15 across the disk radius, and the broadening of the absorption profile is then proportional to length in excess of this threshold (Elitzur, 1992). The saturation issue is not treated in this study.

By comparing the maser brightness temperature T_b at $\theta = 0$ with T_b of the background continuum (10,000–15,000 K; Campbell, 1984), we find that $\tau_0 =$ 18.32 and 15.99 for the 6.7 and 12.2 GHz masers, respectively. We parameterize the function $\eta(\rho)$ as $\propto \rho^{(-p)}$ and set $p = -0.5$, being the value that gives the best fits. The adopted distance to the source D is 2.7 kpc.

2.1. Fitting

The free parameters in the fit are the angular velocity Ω_m and the ratio between inner and outer radii, h. We define $\Omega_m = 1/2(\Omega_i + \Omega_o)$, i.e. the mean of angular velocities at the inner and outer radii, as the angular velocity measured in the data (see straight slope in Figure 3). Because of the natural *scalefreeness* of the model, we have to give an estimate for the scale of the disk, e.g. the outer radius. We choose a value which corresponds to a central mass of 30 $M_\odot$, relying on previous estimates in NGC7538-IRS1 N (e.g. Campbell, 1984). It is to be noticed, however, that the model can produce equivalently good fits when this estimate is not taken into account. The model shows a lower limit at disk sizes corresponding to central objects of subsolar masses. It still performs well when setting the central object to 100 $M_\odot$, although then the outliers at 6.7 GHz cannot be part of the disk.

We fit the position–velocity diagram of the 12.2 and 6.7 GHz methanol maser data observed using VLBI (the latter without outliers, see Figure 1). The 12.2 GHz data fit is considered the *reference fit* because of the higher spatial resolution in the data.

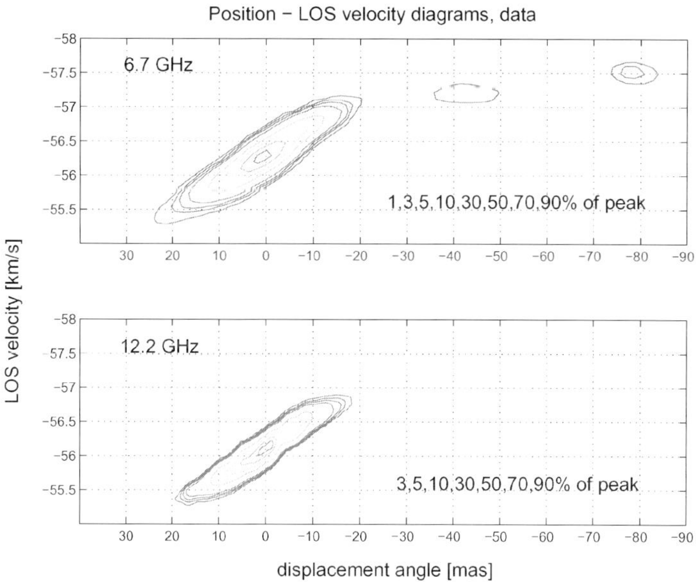

Figure 1. Position–velocity diagrams of the 6.7 GHz (top) and 12.2 GHz (bottom) maser data of NGC7538-IRS1 N . The outliers visible in the 6.7 GHz data were not considered for the fitting.

2.2. The kink

The most important dynamical feature of the model is surely the bend in angular velocity appearing at $|\theta_k| \approx$ 17 mas, visible only in the position–velocity data at 12.2 GHz. This is naturally reproduced by our model and it can be explained in the following way. When $|\theta| \leq \theta_k$ the coherence path participating to the maser extends over the entire disk, leading to the highest τ and so the brightest maser emission. When increasing the displacement above θ_k, the longest coherence path will be centered at velocities closer to the outer edge of the disk, making the increase in angular velocity with displacement angle (or velocity gradient) less pronounced. This will produce a clear bend in the line tracing the maximum optical depth τ at every value of θ. The value for θ_k gets smaller and the bend stronger the smaller $\Delta v_{\rm D}$ or the smaller h. The presence of the bend is a unique signature for a differentially rotating disk (e.g. a Keplerian disk as the one we model here), since a solid-body rotation would show a linear velocity gradient. We consider the independent observation of the bend in data and model to be the most compelling evidence for a circumstellar rotating disk in NGC7538 IRS1 N.

Because of the high requirements in sensitivity and spatial resolution, as well as on the smooth distribution of the masing material in the disk, this bend was not seen when modeling the H_2O maser disk in NGC4258 (Watson and Wallin, 1994). It has been detected in thermal line observations and modeled with the help of multiple dynamical components (e.g. Belloche et al., 2002). Our modeling confirms that there is no need for multiple dynamical components when trying to fit the bend.

3. Results

The results of the fitting of the 12.2 GHz maser data is shown in Figures 2 and 3, $\Omega_m = 0.055$ and $h = 0.36$. These values correspond to a disk where the masing methanol is confined between 270 and 750 AU from the central protostar, which has a mass of $30M_\odot$. The model can reproduce with high accuracy both the brightness profile and the spectrum of the maser emission as well as the indications for a bend in the velocity gradient. The fitting of the 6.7 GHz data gave similar results, i.e. $\Omega_m = 0.055$ and $h = 0.36$. Following the line of maximum τ in the

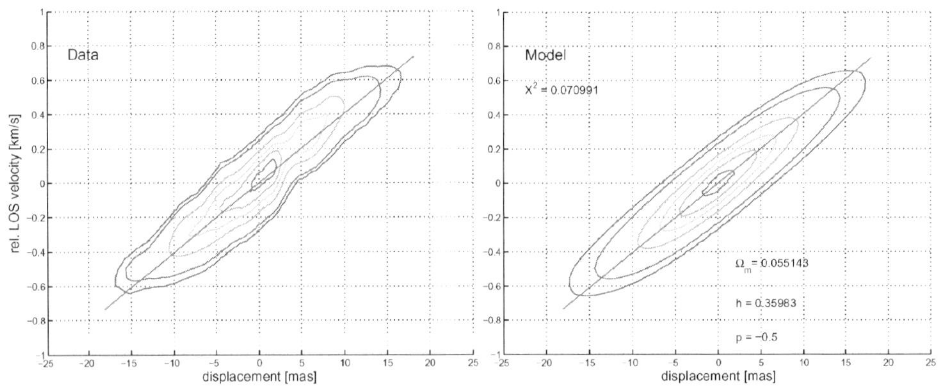

Figure 2. Data (left) and model (right) of the position–velocity diagram of the optical depth τ of the main spectral feature of NGC7538 IRS1 N at 12.2 GHz. In the right panel, the values for Ω_m and h as well as an estimate of the goodness of the fit (X^2) are listed. The blue slope in both panels represents Ω_m.

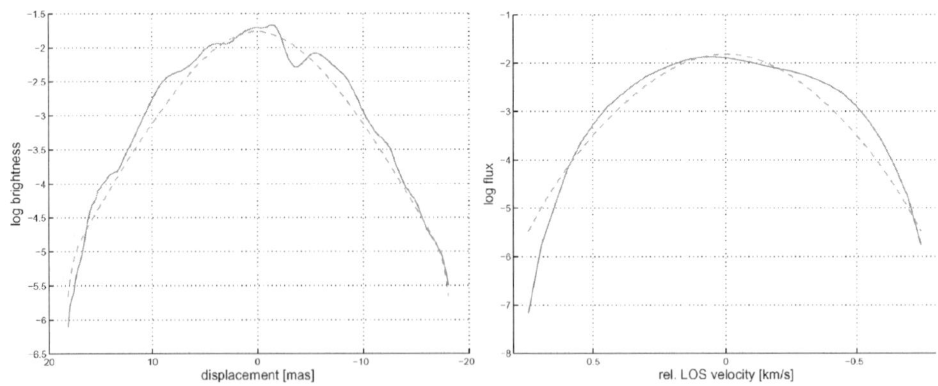

Figure 3. Brightness profile (left) and spectrum (right) of the main spectral feature of the methanol maser of NGC7538 IRS1 N at 12.2 GHz. Filled blue lines are the data, dashed red lines the model.

position–velocity diagram in the adopted model, we can consider the outliers at 6.7 GHz to be part of the disk.

4. Final Remarks

The present study has produced the up to date most accurate model of a circumstellar disk traced by maser emission at small scales ($\lesssim 1000$ AU) around a massive protostar. NGC7538-IRS1 N seems to be a special case, where the masing material appears to be distributed extremely smoothly in the disk and not to be particularly turbulent.

The existence of the outliers in the 6.7 GHz data however indicates that the disk might have a more clumpy structure when approaching the outer edge: the gravitational potential being weaker it is possible to imagine the disk to get disrupted giving rise to local random enhancements of the optical depth offering suitable conditions for maser amplification. In fact, the outliers seem to be very close to the line of highest optical depth in an edge-on disk predicted by our model.

A further issue is to decide whether the background continuum amplified by the inverted methanol in the disk comes from the known UCHII region in NGC7538 IRS1 N and illuminates the entire disk or from the central object solely. The latter case seems to be in accordance with photoevaporation studies, where the radio continuum would be responsible for both the methanol injection into the gas phase and the detected radio continuum outflow (Hollenbach et al., 1994). At the moment, it is not possible to discriminate both scenarios.

Finally, it is important to note that the present study of maser emission does not differ very much from more frequent studies of (optically thin) thermal line emission. Once taken, the logarithm of the maser brightness map the problem reduces to the study of the value of the optical depth τ, which is exactly the same as in the case of thermal emission. The advantage of maser emission is the possibility to achieve high spatial resolution and get information about the dynamics at very small scales (e.g. 1 AU at 1 kpc).

Acknowledgements

M. Pestalozzi thanks R. Parra for the Matlab tricks. We acknowledge A. Jerkstrand for spotting a mistake in the original fitting program. The 12 GHz data cube was kindly made available by V. Minier.

References

Belloche, A., André, P., Despois, D. and Blinder, S.: 2002, *A&A* **393**, 927.
Campbell, B.: 1984, *ApJ* **282**, L27.

Cesaroni, R., Felli, M., Jenness, T., Neri, L., Olmi, L., Robberto, M., Testi, L. and Walmsley, C.: 1999, *A&A* **345**, 949.
Elitzur, M.: 1992, *ARA&A* **30**, 75.
Fuente, A., Rodriguez-Franco, A., Testi, L., Natta, A., Bachiller, R. and Neri, R.: 2003, *ApJL* **598**, L153.
Hollenbach, D., Johnstone, D., Lizano, S. and Shu, F.: 1994, *ApJ* **428**, 654.
Minier, V., Booth, R. and Conway, J.: 1998, *A&A* **336**, L5.
Minier, V., Booth, R. and Conway, J.: 2000, *A&A* **362**, 1093.
Minier, V., Conway, J. and Booth, R.: 2001, *A&A* **369**, 278.
Norris, R., Byleveld, S., Diamond, P., et al.: 1998, *ApJ* **508**, 275.
Qi, C., Kessler, J., Koerner, D., Sargent, A. and Blake, G.: 2003, *ApJ* **597**, 986.
Sandel, G., Wright, M. and Forster, J.: 2003, *ApJ* **590**, L45.
Watson, W. and Wallin, B.: 1994, *ApJ* **432**, L35.

DIAGNOSTICS OF AGN/STARBURSTS

DENSE MOLECULAR GAS AROUND PROTOSTARS AND IN GALACTIC NUCLEI: PDRS AND XDRS

MARCO SPAANS and ROWIN MEIJERINK
Kapteyn Astronomical Institute and Leiden Observatory, the Netherlands;
E-mail: spaans@astro.rug.nl

(Received 16 April 2004; accepted 15 June 2004)

Abstract. The chemical and thermal structure of dense gas irradiated by FUV and X-ray photons is explored. The relevance of these so-called PDRs (FUV) and XDRs (X-ray) for AGN and YSOs is discussed. Observational line diagnostics and chemical tracers are summarized.

Keywords: galaxies: ISM, galaxies: AGN, infrared: ISM: lines and continuum, ISM: atoms and molecules

1. Introduction

Photon-dominated regions (PDRs) are regions in the ISM where the radiation field of nearby stars completely determines the temperature and chemical composition of interstellar gas. As such, PDRs are direct manifestations of the energy balance of interstellar gas and their study allows one to determine how the ISM survives the presence of the stars that it forms (Tielens and Hollenbach, 1985; Boland and de Jong, 1982; van Dishoeck and Black, 1988; Le Bourlot et al., 1993; Wolfire et al., 1993; Spaans et al., 1994; Sternberg and Dalgarno, 1995; Störzer et al., 1995; Spaans, 1996; Bertoldi and Draine, 1996; Maloney et al., 1996; Lee et al., 1996; Kaufman et al., 1999; Le Petit et al., 2002; Tielens and Hollenbach, 1999 and references therein).

Interstellar gas irradiated by optical, UV and X-ray photons from nearby sources is heated, through photo-electric emission from dust grains, (partially) ionized and dissociated. Fine-structure lines of [OI] and [CII] as well as molecular emission of, e.g., CO and H_2O provides cooling. PDRs are bright sources of infrared dust emission because the bulk of the impinging FUV (6–13.6 eV) flux is absorbed by dust grains and re-radiated at longer wavelengths. PDRs are the origins of most of the non-stellar infrared and (sub-)millimeter CO emission from galaxies.

PDRs are found close to bright O and B stars as well as cooler A and F stars. PDRs are also associated with planetary nebulae, active galactic nuclei (AGN) and photo-evaporating planetary disks around newly formed stars as well as diffuse clouds, the neutral inter-cloud medium and molecular clouds exposed to the average interstellar radiation field. Hence, PDRs cover a wide range of astrophysical environments and trace the overall energy balance of the atomic and molecular ISM over ranges in density and temperature of $n_H = 10^2$–10^7 cm^{-3} and $T = 10$–10^4 K, respectively. The term 'PDR' is usually reserved for impinging radiation fields that peak in the FUV, while 'cool PDRs' are associated with stars of effective temperatures of $T_{eff} = 6000$–$20{,}000$ K and 'XDRs' (X-ray dominated regions) are formed near strong sources of X-ray emission.

PDRs and XDRs have become increasingly important as diagnostic tools of astrophysical environments with the advent of infrared and (sub-)millimeter telescopes and space observatories. PDRs emit fine-structure lines of [CI] 609, [CII] 158 and [OI] 63 μm; rotational lines of CO; ro-vibrational and pure rotational lines of H_2; many H_2O lines as well as many broad mid-IR features associated with polycyclic aromatic hydrocarbons (PAHs). In PDRs the bulk of the atomic hydrogen is converted into H_2 and the carbon is transformed from mostly ionized form into CO. XDRs emit brightly in the [OI] 63, [CII] 158, [SiII] 35 and the [FeII] 1.26, 1.64 μm lines as well the 2 μm ro-vibrational H_2 transitions. The abundance of neutral carbon in XDRs is elevated compared to that in PDRs and the chemical transitions from H to H_2 and C^+ to C and CO are smoother.

Detailed models have been constructed of PDRs and XDRs over the past decades and these have allowed us to determine, with increasing confidence: the density and temperature structure, elemental abundances (i.e., the metallicity), the ionization degree of the ambient medium as well as the strength of the impinging radiation field (see Tielens and Hollenbach, 1999; Maloney et al., 1996 for reviews). Various physical processes that play a role in PDRs explain the observed correlation between the [CII] 158 μm (ionized) and CO 1-0 (molecular) emission lines; the link between CO 1-0 emission and the total molecular (H_2) mass, i.e., the X-factor; and also the scaling of the sum of the [OI] 63 μm and the [CII] 158 μm emission (which is almost equal to the total PDR cooling rate) and the infrared continuum luminosity of dust grains.

On a galactic scale PDRs lead to the existence of two stable phases in the ISM, linked to specific temperature thresholds of the atomic coolants, that are roughly in pressure equilibrium. The warm (mostly) neutral medium with a density of $n_H \sim 1$–10 cm^{-3} and a temperature of $T \sim 10^4$–10^3 K and the cold neutral medium with $n_H \sim 100$ cm^{-3} and $T \sim 100$ K (Field et al., 1969). The hot (10^6 K) interiors of supernova explosions form a third phase while the cold (10 K) molecular gas out of which stars are ultimately born is formed under the influence of gravitational collapse of dense molecular clouds.

Finally, PDRs and XDRs are regions where molecular material, necessary for the continued formation of (proto-)stars, is destroyed by radiation from already formed stars. As such, they are regions of strong radiative feedback that plays a part in the regulation of the overall star formation rate from individual molecular clouds to entire galaxies by limiting the column of molecular gas that is available for gravitational collapse (McKee, 1989).

2. Background Physics

2.1. PDRs

A number of basic physical processes play a role in the global properties of PDRs: (1) Heating through photo-electric emission by dust grains and PAHs (c.f. Bakes and Tielens, 1994). (2) Heating by cosmic rays (c.f. Field et al., 1969). (3) Fine-structure line cooling of [CI] 609, [CII] 158, and [OI] 146 and 63 μm (c.f. Tielens and Hollenbach, 1985; Spaans et al., 1994). (4) Molecular line cooling by warm molecular gas containing CO, H_2, H_2O, OH and CH (c.f. Neufeld et al., 1995; Spaans and Silk, 2000). (5) Ion-molecule reactions driven by the ionization degree of $\sim 10^{-4}$ maintained by ionization of carbon in the FUV (c.f. Black and Dalgarno, 1977; van Dishoeck and Black, 1986). (6) The ionization balance of atomic gas under the influence of photo-ionization reactions driven by FUV photons and counteracting recombination and charge transfer reactions with metals and particularly PAHs (c.f. Lepp and Dalgarno, 1988; Bakes and Tielens, 1998).

As one moves into a PDR the extinction along the line of sight increases and the impinging radiation field is attenuated. Consequently, there are two zones over which the chemical composition of the PDR changes in a fundamental way. The first fundamental change occurs at the very edge of the PDR as atomic hydrogen is converted into H_2 because the Lyman and Werner electronic bands that lead to dissociation of the H_2 molecule in the FUV become very optically thick (so-called self-shielding). Deeper into the PDR, at about 3 mag of extinction, ionized carbon is quickly converted into neutral form as the FUV flux deceases due to dust absorption and C is subsequently transformed into CO because the latter molecule also self-shields.

The first few magnitudes of extinction of the PDR are usually referred to as the radical region since many carbon hydrides and their ions, e.g., CH, CH^+, CN, HCN, HCO^+ (and also CO^+), reach their peak abundance there, caused by the presence of both C^+ and H_2 and the high ($\sim 10^2$–10^3 K) temperatures. Ion-molecule reactions take place that lead to the formation of a large number of different molecular species. Many of the atoms and molecules in (the radical region of) a PDR are collisionally excited at the ambient densities and temperatures, and emit brightly in the mid-IR, FIR, millimeter and sub-millimeter.

The global characteristics of any PDR are defined by a few key parameters. (a) The strength of the impinging radiation field, G_0 or $I_{\rm UV}$, in units of the Habing (1969) or Draine (1978) radiation field, respectively, determines the total available radiative flux at the edge of the PDR. (b) The ambient hydrogen density, $n_{\rm H} = n({\rm H}) + 2n({\rm H_2})$, sets to a large extent (together with the temperature) the pace of chemical reactions and the excitation rate of coolants. (c) The metallicity Z, in units of the Solar value $Z_{\odot}$, constrains the total abundances possible for carbon- and oxygen-bearing species and hence influences the chemical and thermal structure. (d) The spectral shape of the impinging radiation, parameterized by the color temperature $T_{\rm eff}$ for black bodies or the frequency slope for power laws, fixes the distribution of photon flux over energy.

2.2. XDRs

In XDRs heating is provided by photo-ionization, which produces fast electrons that loose energy through collisions with other electrons as well as H and H_2. These fast electrons cause collisional excitation of H and H_2, which subsequently emit Lyman α and Lyman-Werner band photons, respectively. The latter photons are capable of ionizing atoms like C and Si or ionize/dissociate molecules like H_2 and CO.

Similarly to PDRs, the following processes play a role in XDRs (c.f. Maloney et al., 1996), in part because of the production of UV photons as described above. (1) Photo-ionization heating (i.e., Coulomb heating with thermal electrons) that dominates by a large factor over heating through photo-electric emission by dust grains and PAHs (c.f. Maloney et al., 1996; Bakes and Tielens, 1994). (2) Emission from meta-stable lines of [CI] 9823, 9850 Å and [OI] 6300 Å; fine-structure line cooling of [CII] 158 and [OI] 146 and 63 μm as well as Lyman α emission (c.f. Maloney et al., 1996; Tielens and Hollenbach, 1985; Spaans et al., 1994). (3) Molecular line cooling by warm molecular gas containing CO, H_2, H_2O and OH as well as gas-grain cooling where warm gas is cooled on the surfaces of lower temperature dust grains (c.f. Neufeld et al., 1995; Spaans and Silk, 2000). (4) Ion-molecule reactions driven by the ionization degree of $\sim 10^{-4}$ maintained by ionization of carbon in the FUV (c.f. Black and Dalgarno, 1977; van Dishoeck and Black, 1986). (5) The ionization balance of atomic gas under the influence of photo-ionization reactions driven by X-ray photons and charge transfer. Note that recombination of ions on grain surfaces can become a major ionic loss route once the electron fraction becomes less than 10^{-3} (c.f. Lepp and Dalgarno, 1988; Bakes and Tielens, 1998; Maloney et al., 1996).

The global characteristics of any XDR are also defined by a few key parameters, the density $n_{\rm H}$ and the energy deposition rate $H_X/n_{\rm H}$. The latter measures the amount of X-ray energy that is deposited into the medium per hydrogen atom. Note that because the heating in XDRs is driven by photo-ionization, the heating efficiency is close to unity whereas it is of the order 0.3–1.0% for photo-electric heating in PDRs (Maloney et al., 1996; Bakes and Tielens, 1994). Unlike PDRs, XDRs are exposed to X-ray and FUV photons.

As one moves into the XDR, X-ray photons are attenuated for a column density of more than $\sim 10^{22}$ cm^{-2}. The attenuation is a consequence of atomic electronic absorptions. With increasing column density, the lowest energy photons are attenuated strongest, which leads to a dependence of the X-ray heating and ionization rates on the slope of the X-ray spectrum. For AGNs, one typically has a power-law index $\alpha \approx 1$, i.e., constant energy per decade. One may assume, for energies between 0.1 and 10 keV, that the primary ionization rate of hydrogen is negligible compared to the secondary ionization rate and that Auger electrons contribute an energy that is equal to the photo-ionization threshold energy (Voit, 1991).

3. Results

Figures 1–8 show the general characteristics of PDRs and XDRs. For the XDRs we consider a column of 10^{22} cm^{-2} so that attenuation effects of the incident X-ray flux can be ignored. However, variations in the energy deposition parameter H_X/n_H mimic attenuation effects for a given hydrogen density quite well. Note that the code used to generate these figures treats PDR and XDR physics simultaneously.

For the PDR model the density and strength of the radiation field, $G_0 = 10^5$ and $n_H = 10^5$ cm^{-3} are typical of the Orion Bar and meant to illustrate the effects of feedback of high-mass star formation on the parental molecular cloud (Tielens and Hollenbach, 1985). For starburst galaxies (like NGC 253) star formation rates of $\sim 10^2 M_\odot$/yr should lead to similar conditions in their dense central regions.

For the XDR model we consider a range in the energy deposition parameter H_X/n_H with $n_H = 10^5$ cm^{-3}. This range covers both Seyfert galaxies (typical X-ray luminosity of $L = 10^{44}$ erg/s) as well as T-Tauri stars (typical X-ray luminosity of 10^{32} erg/s) given that the impinging X-ray flux is given by $F_X = 84L_{44}/r_2^2$ erg cm^{-2} s^{-1}, where L_{44} is in units of L and r_2 in units of 100 pc (Maloney et al., 1996). AGN (like NGC 1068) gas is typically located at distances $r = 0.1$–1 pc while this is $r = 20$ AU for YSOs. Both PDR and XDR models are for a slab geometry.

Figure 1 shows for the PDR the photo-electric emission as a function of extinction. At the edge of the cloud the temperature is about three times higher when

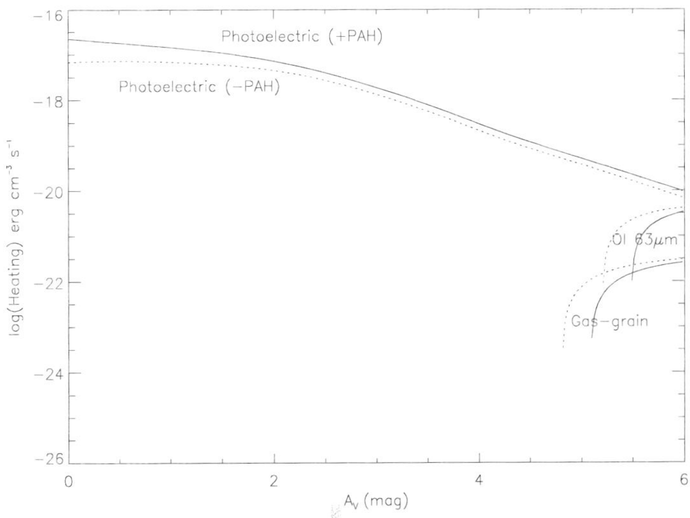

Figure 1. Heating processes in a PDR with (solid line) and without (dotted line) PAHs.

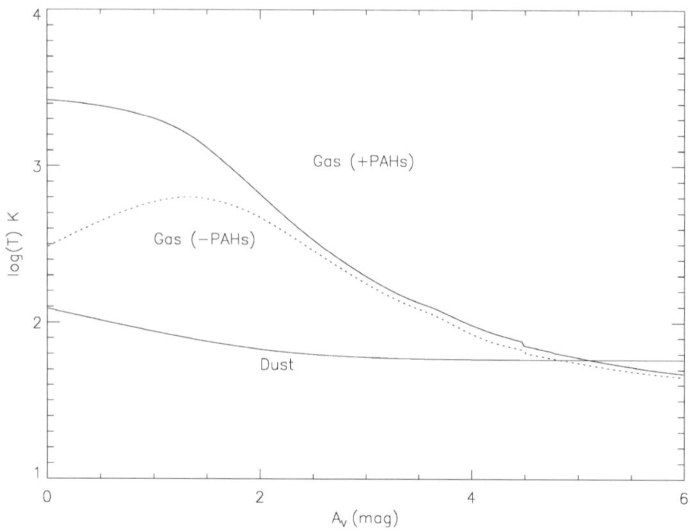

Figure 2. The temperature with (solid line) and without (dotted line) PAHs.

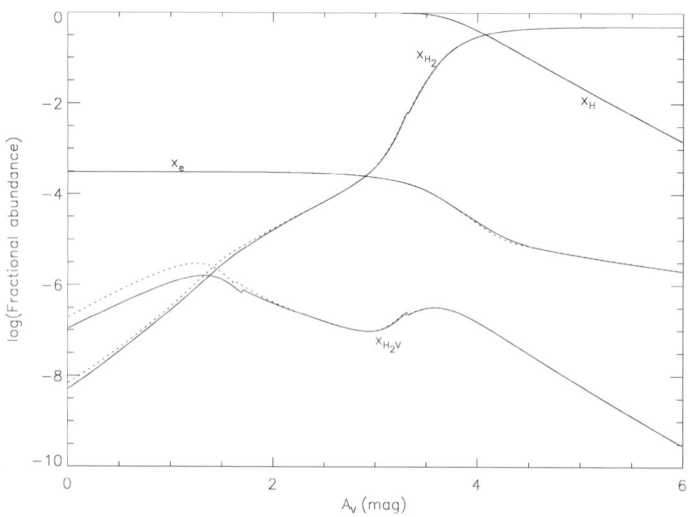

Figure 3. The H/H_2 transition in a PDR, with (solid line) and without (dotted line) PAHs.

PAHs are included (see Figure 2). Deeper into the PDR fine-structure and gas-grain heating become important.

In a XDR, the photo-ionization heating dominates photo-electric heating by two orders of magnitude. As mentioned above, X-ray photons create fast electrons through photo-ionization. About 40% of their energy is lost through Coulomb interactions with thermal electrons.

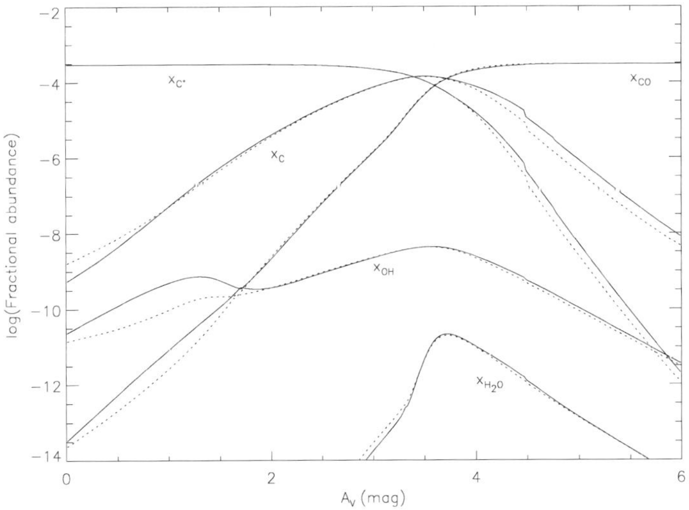

Figure 4. The C^+/C/CO transition in PDR, with (solid line) and without (dotted line) PAHs. Abundances of H_2O and OH are also shown.

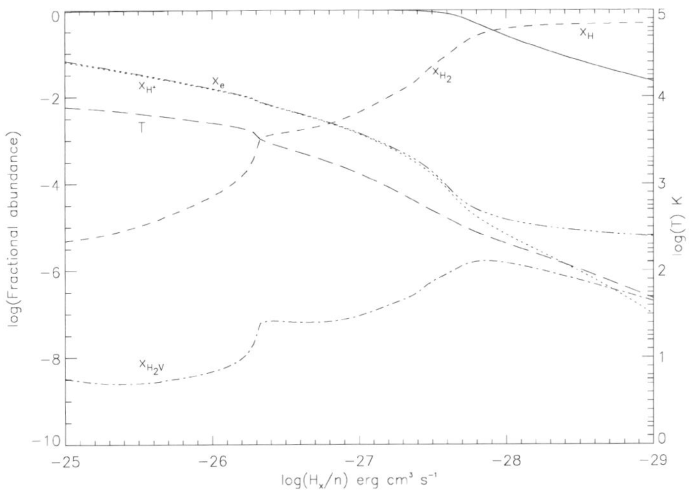

Figure 5. The abundances of hydrogen species, H^+, H, H_2, H_2V, the elcctron abundance and temperature T.

In a PDR, the chemical structure is stratified. The FUV photons are gradually absorbed and lead to relatively sharp transitions. In Figure 3 the transition of atomic to molecular hydrogen is shown. Since the temperature is larger when PAHs and small grains are included, the H_2 abundance between 1 and 2 mag is somewhat smaller due to dissociative collisions. Molecular hydrogen is excited by fluorescence (i.e., absorption of FUV photons). In Figure 3 this is shown as H_2V, which indicates a $v = 6$ pseudolevel of H_2 at about 2.6 eV above ground level. In Figure 4

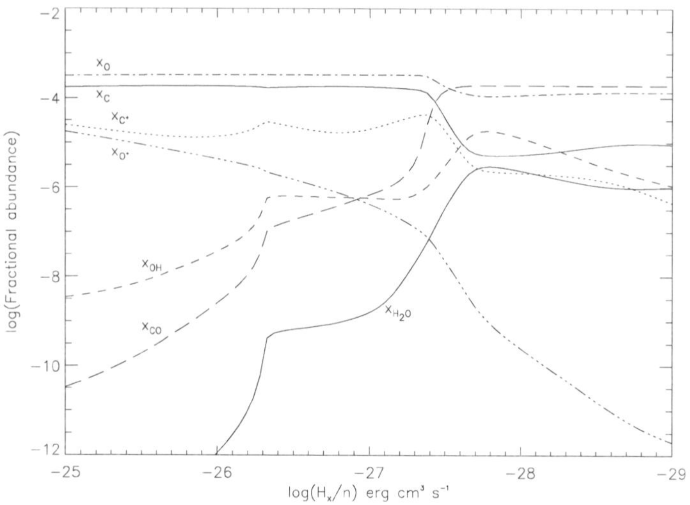

Figure 6. The abundances of O, C^+, C, CO, H_2O and OH.

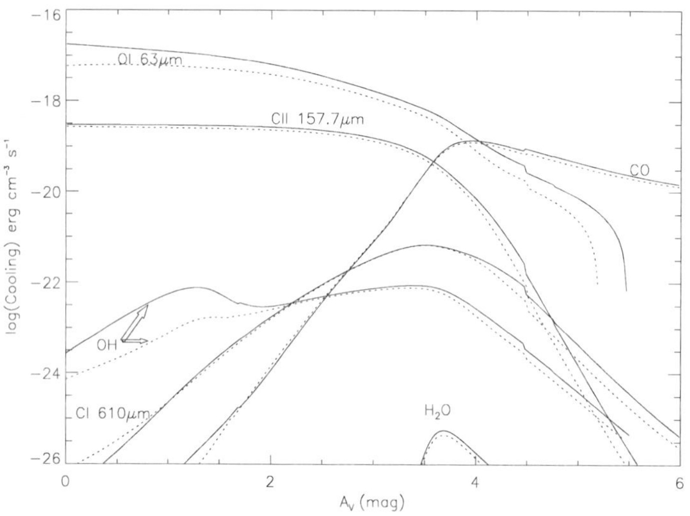

Figure 7. Important coolants in a PDR.

the transition C^+/C/CO is shown. Products from endoergic reactions like $O+H_2$, leading to OH and H_2O, are more abundant at the warm edge of the PDR when PAHs are included.

X-rays can penetrate deep into a cloud and the chemistry is not stratified. We show the chemistry as a function of X-ray energy available per particle at a typical attenuation column of 10^{22} cm^{-2} (see Figures 5 and 6). For large X-ray energies per particle, the electron abundance is high, due to the ionization of hydrogen by secondary electrons. C^+ is always present but never dominant.

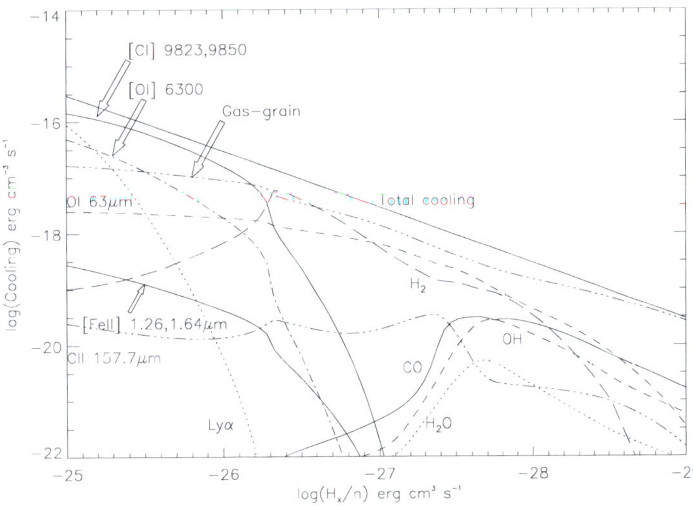

Figure 8. Important coolants in a XDR.

At high X-ray energies, the only way to produce C^+ is by primary and secondary ionizations, of which the rates are comparable. Lyα photons, produced by the excitation of atomic hydrogen, cannot ionize carbon. At smaller X-ray energies, Lyman-Werner photons dominate the ionization of carbon. H_2 is excited due to collisions with electrons. H_2O and OH are more easily formed when H_2V is present since the 2.6 eV of internal energy is sufficient to overcome reaction barriers in the oxygen chemistry. Hence, molecular and ionized species coexist in a XDR.

At the edge of a cloud in a PDR the dominant coolant is the [OI] 63 μm fine-structure line (see Figure 7). The [CII] 158 μm/[OI] 63 μm ratio is sensitive to the density distribution and the temperature. Deeper into the cloud cooling by CO becomes dominant. Cooling by H_2O and OH are also shown.

A XDR is highly ionized when a large amount of X-ray energy is available. Thermal electrons can excite forbidden lines like [CI] 9823, 9850 Å and [OI] 6300 Å , but also the [OI] 63 μm line is very strong. The [FeII] 1.26 and 1.64 μm lines can also be very useful tracers when the X-ray illumination (i.e., the temperature) and (column) density is high. When less X-ray energy is available H_2 and gas-grain cooling are important. CO, OH and H_2O are never dominant coolants.

Although not shown, a correlation should exist between CH and HCN since both are formed efficiently in the radical region of a PDR or XDR where C^+ and H_2 co-exist. Finally, hard X-rays, but also supernova-driven shock waves, are capable of destroying small dust grains (Krolik, 1997, private communication), which leads to an increase in the total dust surface area and hence to a larger H_2 formation rate and ionic recombination rate on dust grains.

2.3. OH-MM in Mrk 231

The AGN-dominated ULIRG galaxy Mrk 231 shows clear evidence of past interaction and is optically classified as a Seyfert 1. The active nucleus exhibits a radio jet at 3 pc at PA = 60° that rotates to PA = 6° out to a distance of 100 pc (Ulvestad et al., 1999). On the other hand, there is extended radio emission at scales of some 300 pc and an arc of star-formation at 2 arcsec to the SE. At a distance of 172 Mpc for Mrk 273, 1 arcsec corresponds to 835 pc.

VLBI measurements with the EVN at 35 mas = 27 pc show that the OH emission in Mrk 231 extends on a scale of 170 mas, and lies partially on top of nuclear radio sources combined with the radio outflow structure (Klöckner et al., 2003). The spatial structures of the OH emission and the radio continuum have been presented in Figure 1a. These measurements recover only about 50% of the single-dish flux density is missing on scales between MERLIN resolution (Richards et al., 2000) and WSRT resolution. The OH signature suggests moderate amplification gain for the emission in the range of 2–10. The line ratios vary from the LTE value of 1.8 to 12. On the SW side of the emission structure, there is a region with exceptionally broad emission lines and most extreme values for the line ratio. This region could be affected by intense shocks and interaction with the nuclear jet. The OH emission data in this AGN powered source has been interpreted in terms of a tilted torus structure (Figure 1b). The molecular material in this torus extends across about 130 pc and has an inner radius of 40 pc. The torus structure and the detected

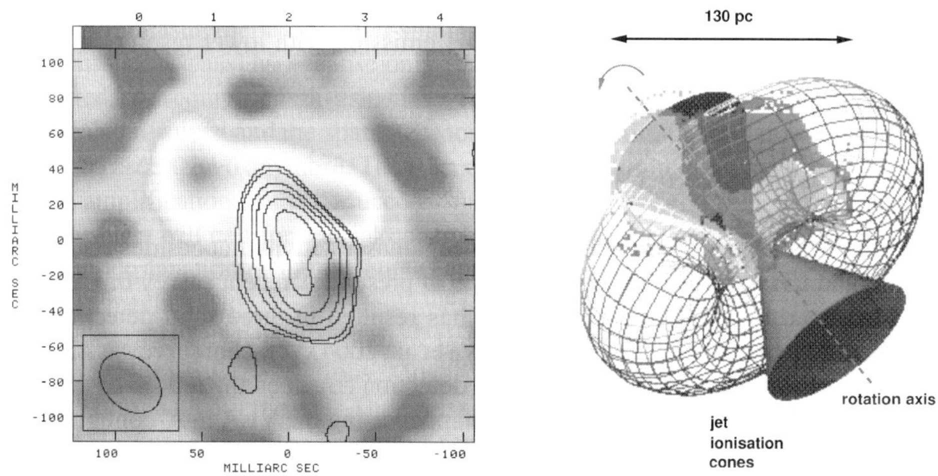

Figure 1. The emission characteristics of the OH Megamaser emission in Mrk 231 observed with the European VLBI Network. (a) The grey-scale depicts the extent of the line emission while the contours depict the compact radio emission of the AGN plus jet. (b) Modeling results for a tilted torus structure that fits the available parameters for the nuclear region of Mrk 231. Figures taken from Klöckner et al. (2003).

velocity gradient suggest a central BH mass of 7.2×10^7 $M_\odot$ (Klöckner et al., 2003).

2.4. OH-MM in Mrk 273

The merging galaxy Mrk 273 displays a complex optical structure with dust lanes, and contains multiple nuclei optically classified as SBN and AGN. This multiple structure is also found in the radio (Carilli et al., 2000). At 154 Mpc for Mrk 273, 1 arcsec corresponds to 745 pc.

The OH emission has recently been studied with the EVN at 47 mas = 35 pc resolution by Klöckner and Baan (2004). The OH emission spectrum of Mrk 273 displays a triple signature, which is seen not only in single-dish data but also at VLBI resolution, where 60% of the mostly broad OH emission is missing. The OH emission is found partially on top of the radio continuum of the Eastern of the two prominent components in the northern part of the source (see Carilli et al., 2000; Xia et al., 2002). The OH emission has a spatial extent of 180 pc and is found to be partially on top of the high brightness continuum source found with VLBI. The velocity contours of the OH emission have been displayed in Figure 2a, while the continuum structure is given in Figure 2b. The displacement of the line emission and continuum is depicted by the North–South spatial offset of the continuum. The line/continuum ratio is moderate across the source with a maximum of 14. The North–South velocity gradient of the OH emission has been interpreted in terms of an edge-on Keplerian disk/torus structure, surrounding a central BH mass of 1.4×10^9 $M_\odot$.

3. The Properties of the Emission Structures

As the fraction of active sources increases with L(FIR), OH Megamaser activity, a signpost of the ULIRG population, may thus serve as a probe for nuclear activity in redshift space. The velocity structure of the molecular lines probes the velocity field of the nuclear region. In the case of absorption and possibly emission, the velocity field lies in front of the nuclear radio emission. The velocity width is composed of different contributions: the separation of the 1665 and 1667 MHz components, rotation in a disk or torus, velocities due to galaxy interactions, possible molecular inflows, and molecular outflows as part of a nuclear superwind.

An early assessment of the line profiles of powerful megamasers shows that outflows are quite common in OH emitters and absorbers (Baan et al., 1989). Outflow velocities up to 800 km/s have been seen and it appears that the velocity extremes vary with the energy generation in the nucleus as exemplified by L(FIR).

A recent high-resolution study of the nearby galaxy NGC 3079 has confirmed the presence of two distinct OH and HI absorption components against the nuclear

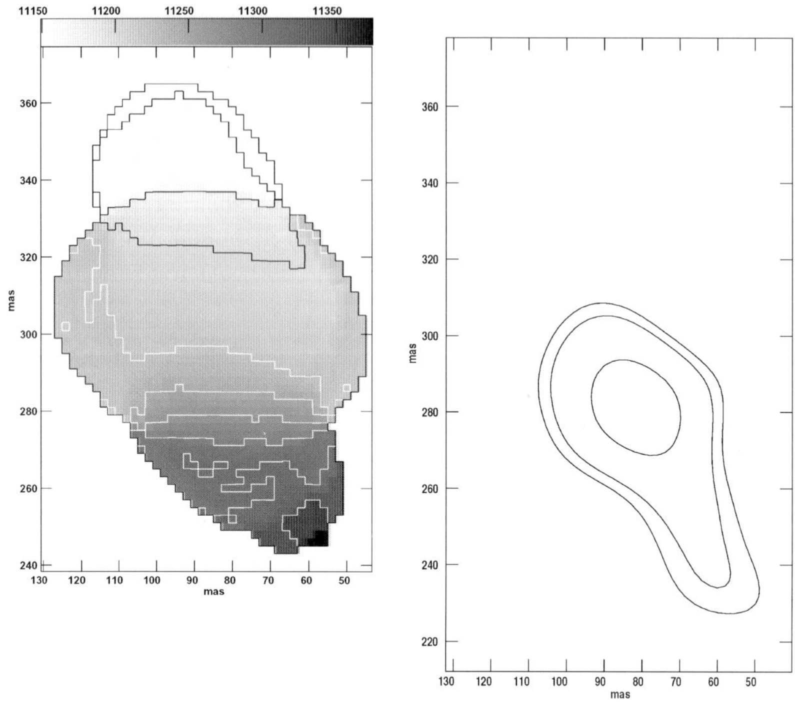

Figure 2. The structure of the OH-MM emission in Mrk 273 detected with the European VLBI Network. (a) The velocity field of the OH emission is depicted in the left frame. The figure clearly shows a strong velocity gradient across the nuclear region. (b) The right frame depicts the relative location and extent of the radio continuum emission on the same scale as in frame (a). The continuum emission covers only a small fraction of the line emission region. Figures taken from Klöckner and Baan (2004).

radio continuum (Hagiwara et al., 2004). While in earlier papers these components were thought to originate along distinct lines of sight toward two radio components in the nucleus, it appears that one absorption component is at a systemic velocity and results from absorption against the whole nucleus. The second component is also in front of the whole nuclear continuum but represents an outflowing molecular shell at -125 kms/s. The velocity gradient in this shell is almost opposite that of the disk gas within the systemic absorption component. This molecular outflow provides more observable evidence of the spectacular nuclear blowout that characterizes the nucleus of NGC 3079. A similar shell-like molecular outflow has been seen in M82, where the shell contains also maser emission (Pedlar et al., 2004; see also these proceedings).

The current sample of OH Megamasers traces the high luminosity FIR galaxy population and continues to a redshift of 0.265. The sources detected at higher z confirm that the isotropic luminosities of the OH-MM can indeed reach well above $L(\mathrm{OH}) = 2 \times 10^4\ L_\odot$. However, the examples of such distant high-luminosity sources are characterized by extremely large total velocity widths of

TABLE I
Torus characteristics

	Torus outer size	V(rot)	Central Mass
IIIZw35	86 pc	65 km/s	$7 \times 10^6\ M_{\odot}$
Mrk 231	200 pc	140 km/s	$7.2 \times 10^7\ M_{\odot}$
Mrk 273	108 pc	118 km/s	$1.4 \times 10^9\ M_{\odot}$

up to 2000 km s^{-1}. The most extreme examples of such sources are IR14070 + 0525 ($z = 0.265$; Baan et al., 1992) and IR12032 + 1707 ($z = 0.217$; Darling and Giovanelli, 2001). VLBI studies of these two sources show their complexity and their emission profile make it difficult to disentangle the various velocity components (Pihlström et al., 2004).

The high-resolution studies in IIIZw35 (Pihlström et al., 2001), Mrk 231 and Mrk 273 show a consistent picture for the torus structure that contains the OH emitting gas. The outer diameters found for the tori are in the range of 86 to 200 pc. The rotational velocities vary between 65 and 140 km/s depending on the central mass.

The inclination of the dusty torus affects also the FIR colours and particularly the 60 μm to 25 μm IR colour (Heisler et al., 1997). Highly inclined (edge-on) sources will have smaller values of the 25 μm/60 μm ratio. On the other hand the naked nuclei of Seyfert 1's have much larger ratios.

4. Molecular Densities in Megamaser Sources

X-ray studies show that the absorbing column density of cold absorbers and scatterers in Seyferts vary over several orders of magnitude (see Xia et al., 2002). Face-on systems in Seyfert 1's have little or no absorption and $20 < \log(N_H) < 21$, while the edge-on Seyfert 2's have significant to extreme absorption with $22 < \log(N_H) < 25$. Converting a column density of $\log(N_H) = 24$ in a torus with a 70 pc cross section, one finds a mean density of 4.6×10^3 cm^{-3}.

The column densities required for OH and formaldehyde emission can be determined from the optical depth in the maser lines and can be compared to the nominal H column density seen in an edge-on torus in Seyferts of $\log(N_H) = 24$. Assuming $T_{ex} = 10$ for both H_2CO and OH, the OH column density estimates for the five sources discussed above vary between 6×10^{17} and 2×10^{18} cm^{-2} and have an OH/H abundance close to the Galactic range of 10^{-6}–10^{-7}. Similarly the H_2CO column density of the two formaldehyde masers gives a column density of 2 and 4×10^{15} cm^{-2} and an H_2CO/H ratio of 2–4 $\times 10^{-9}$, which is higher than the Galactic value of 10^{-10}. The preliminary numbers found for these megamasers suggest approximate galactic molecular parameters for the torus material that makes up the megamaser emission regions.

5. High Density Tracer Molecules

High-density molecular gas plays a crucial role in the physics of (U)LIRGs, giving rise to spectacular starbursts and possibly also providing the fuel for a highly obscured active galactic nucleus (AGN). Studies of molecular gas in less active nearby galaxies have revealed large amounts of molecular gas in some nuclear regions (Solomon et al., 1992; Hüttemeister et al., 1995). As the L(HCN) – L(FIR) correlation is found to be tighter than the L(CO) – L(FIR) correlation, the high-density gas in the nuclear regions of galaxies appears to have a closer relation to nuclear activity (Solomon et al., 1992). Furthermore, the molecular gas in the central regions of FIR galaxies makes up a significant fraction of the dynamical mass (Henkel et al., 1990; Solomon et al., 1992). In particular, tracers of the high-density components of the molecular medium are indicative of the conditions in these star-formation regions.

Recent studies by Aalto et al. (2002) and Baan et al. (2004) confirm that there is a strong linear correlation between the luminosity of the high-density tracer lines and the FIR luminosity. The normalized line ratios for HCN(1-0), HCN(1-0), and HCO^+(1-0) as compared with the CO(1-0) lines have been presented for a number of ULIRGs and OH-MM in Figure 3. These normalized line ratios show a large spread at each L(FIR), and no systematic behavior has yet been seen within this parameter space. On the other hand, the data points are quite well organized and are located on a wedge in the diagrams starting at high-FIR luminosity and getting wider at lower luminosity. The range for the ratios of HCN and HNC is significantly larger than for HCO^+. Also the slope of the wedge is different for the three molecules. This wedge suggests that there are no sources with high values of the low-density versus high-density ratio at high FIR luminosities, and that the high-density molecular component becomes less prominent for lower luminosity sources.

The molecule HCN is the most stable of the ones considered in Figure 3 and has an abundance that increases with increasing gas deposits. The HCN abundance reflects the stage of chemical evolution of the galaxy. Steady state models show a decrease of the HCN abundance and the HNC/HCN ratio for increasing temperatures (Schilke et al., 1992). The observed line ratios in Figure 3 appear consistent with those of dense gas unaffected by shocks. The two prominent galaxies NGC 6240 and Arp 220, which appear quite similar in their evolutionary state, show a lack of systematic behavior in that they have very different HNC abundances.

The ratio HCO^+/HCN increases when there are high cosmic ray fluxes and high UV fluxes in shocks (Schilke et al., 1992. In addition, a few detections at the nuclear regions have been made for CN, which is a tracer of the chemical age of the galaxies.

In Figure 3 the OH-MM occupy the region with the highest FIR luminosities. The values of the *LHrat* density ratio are either high or low and appear to lie on the outer envelop of the distribution of points. In terms of the evolutionary track of

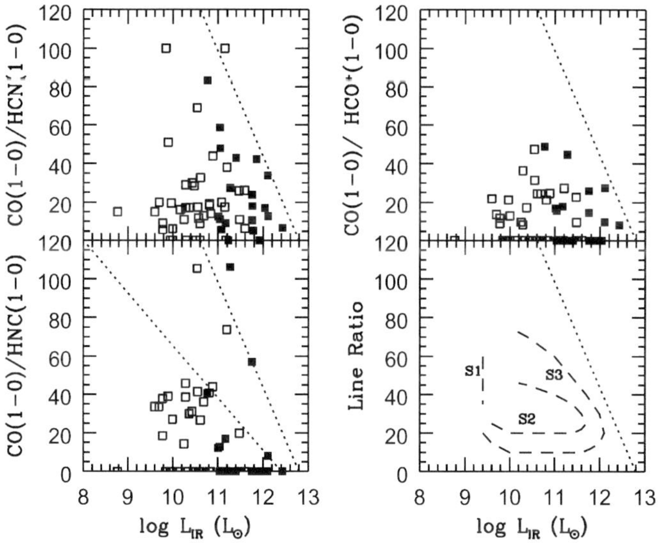

Figure 3. The normalized line ratio diagrams of high-density tracer molecules in ULIRG-OH MM galaxies. The emission line flux for the HCN(1-0), HNC(1-0), AND HCO^+ transitions are normalized with those of the lower-density tracer line of CO(1-0). The solid squares represent megamaser galaxies and the open square are non-megamasers. The bottom right panel shows the three stages S1, S2 and S3 that are part of the evolutionary track of an ULIRG during a period of nuclear activity (Baan et al., 2004).

the ULIRGs in Figure 3d and as discussed in the section below, these points would then lie on ULIRG evolutionary curves that reach the highest FIR luminosities.

6. An Evolutionary Scenario of a ULIRG Outburst

During the course of the luminosity evolution of a starburst dominated ULIRG, the dusty regions responsible for the FIR emission coincide for a large measure with the high-density molecular regions spread throughout the nuclear region, where the active star-formation happens. If the dominant energy source of the ULIRG outburst were an AGN, then the FIR might result from re-emission from a dusty/molecular torus. Similarly, the molecular material in a Starburst Nucleus (SBN) could have a similar torus configuration.

When studying the properties of high-density tracers in the nuclear region, the strong dependence of these tracers on local conditions would require that a comparison be made with the properties of the low-density gas components that trace regions outside the regions of nuclear activity. The molecular gas traced by the high-density tracers is predominantly concentrated toward the nuclear region, and represents the dusty clumps of high-density material that also serve as the environment for the star-formation. On the other hand, it is known that the low-density tracer line emissions, such as the CO(1-0) transition, would predominantly originate in the more diffuse (lower-density) gas spread throughout the surrounding regions in the central and outer galaxy. The environments of the high-density and low-density components would thus be quite different.

The low-density and high-density molecular components of the galaxy will fare quite differently during the course of a burst of star-formation. While the outlying regions containing low-density gas may indeed be partially affected by the star-formation process, the state of its large-scale low-density molecular structure would not be strongly altered. On the other hand, the high-density gas will be strongly affected because it provides the material for the star-formation activity. During the star-formation process the high-density molecular gas will thus be slowly consumed by the newly formed stars.

The FIR luminosity of the ULIRG during its outburst reflects the amount of energy generated by the nuclear activity, which is mostly related to the time-delayed star-formation activity. At the same time, the luminosity of the ULIRG integrated over the course of its outburst would reflect the amount of high-density molecular material consumed and destroyed by the burst of star-formation.

In reality, quite complicated processes govern the emission mechanism of the various molecules, which depend on gas density, gas pressure, ambient radiation fields, and their relative abundance. However, in advance of detailed modeling of the molecular emissions, one may identify the following *three distinct stages* for the history of the dense molecular material in the galaxy (Baan et al., 2004).

6.1. Rise stage (1)

There will be a rise in the FIR luminosity during the early evolution of the ULIRG activity with no significant depletion of the high-density component and only small changes in the low-high line ratio. This stage would result in a shift to the right for a data point in the Figures.

6.2. Decay stage (2)

After the peak in the FIR luminosity there will be a slow decay of the FIR luminosity and a steady depletion/destruction of the high-density component, which will result in an increase in the low-high ratio. This stage will result in a left-upward movement in the Figures.

6.3. Build-up stage (3)

During this stage there will be a slow building up of the high-density molecular component in the central regions of the galaxy. The build-up of the high-density component will result from radial mass-flow inward caused by possible mergers or interactions with neighbors and generally from certain perturbations in the gravitational potential in the galaxy, which could be invoked by nuclear bars. In this stage the build-up will prepare the galaxy for the next outburst.

To simulate these stages in the evolution of the nuclear activity, we make the following assumptions:

(1) the FIR luminosity curve of the nuclear activity is characterized by a rise and fall with different timescales. For simplicity we use the standard expression of a diffusion function with time t defined as: $\mathrm{LFIR}(t) = A(T/t)^{2.5} \exp(-T/t)$, where T is a representative timescale of the outburst, which depends on the size of the region and the propagation speed of the star-formation activity. A is a scaling factor for the magnitude of the outburst. Although such a diffusion curve may not be the most appropriate representation, it serves quite well for our purpose of showing the general dependence of FIR luminosity and content of high-density gas of the galaxy. A Gaussian or a sinusoidal luminosity evolution will give similar results.

(2) the amount of high-density material that disappears as a function of time during the outburst is proportional to the integrated FIR luminosity up to that point in the evolutionary process. At the same time, the amount of low-density material remains approximately the same.

Simulations along these lines have been done and the results obtained have been presented in Figure 4. These simple time-dependent relations give a correct behavior for the ratio *LHrat* of low-density and high-density material in the vertical axis versus L(FIR) in the horizontal axis. The curves stay within the parameter space observed for the various high-density molecules. If the range of *LHrat* covers

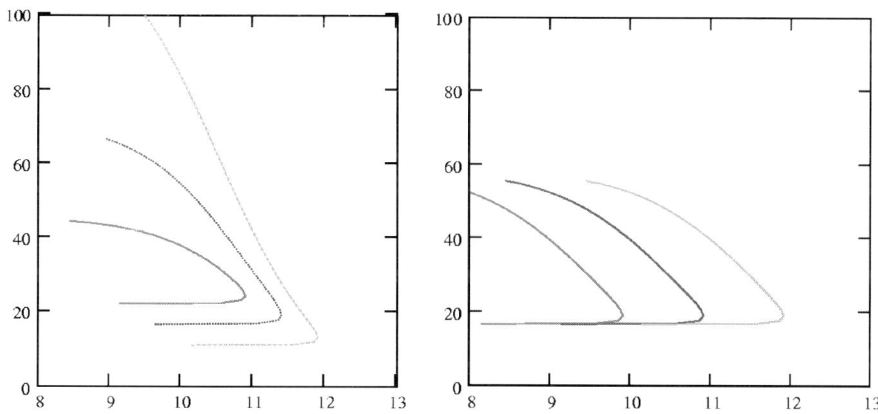

Figure 4. Simulations of the evolutionary history of the high-density molecular component of ULIRGs undergoing periods of intense star-formation. The vertical axis represents the ratio *LHrat* of low-density tracer lines over those of high-density tracers as depicted in Figure 4. The horizontal axis represents the FIR luminosity in $L_{\odot}$. The left curve depicts different initial values for *LHrat*, while the right curve depicts different peak values for L(FIR) with the same initial value of *LHrat*.

the whole range observed in the diagrams of Figure 3, the fraction of high-density depletion during the outburst should be in the 0.6–0.8 range. Similarly, the initial *LHrat* should be in the 0.09–0.03 range. The peak FIR luminosity should lie in the 10^{10}–$10^{12.5}$ $L_\odot$ range. Naturally, the parameterizations presented here are not unique but they do present a realistic and acceptable scenario and pattern for the global processes that could govern the molecular ratios. If the luminosity curve had a more delayed peak than the one chosen, the evolutionary curves would be more rounded and the turn-around point between Stages (1) and (2) would occur at higher *LHrat* values.

It should be noted that the upper envelops of the data points observed in Figure 3 indicate a steady change in *LHrat* during the Decay Stage of the outburst. During this stage of the outburst, our simulated FIR luminosity varies as $\exp(-T/t)$ according to the diffusion equation mentioned above. This would suggest that for the Decay Stage (2) of all the curves, which is the dominant part of the time-history of the outburst, *LHrat* varies with time as (t/T) and that the high-density component varies as (T/t). During the Decay Stage of the outburst, the high-density component of the molecular gas in the nuclear region would decay linearly with time.

7. Concluding Remarks

The OH-Megamasers occupy a special place in the population of ULIRGs and represent an important stage in the evolutionary track of these sources. In terms of the evolutionary track of the ULIRGs, the Megamasers would then only lie on evolutionary curves that reach the highest FIR luminosities. The presence of abundant high-density material is the guarantee for intense star-formation activity and high peak luminosities in the FIR. On the other hand, the presence of high-density molecules is also a measure of the OH column densities in the nuclear region, and of the FIR radiation field that determines the pumping conditions. Thus Megamaser activity and a significant high-density component are strongly related.

The combined behavior of the high-density tracers illuminates the relevant physical processes in the nuclear region, the chemical history, and the relative age of the nuclear activity. Using molecular lines as diagnostic tools for probing the nuclear ISM in Megamaser galaxies and ULIRGs will cover a regime of nuclear activity that has not yet been addressed by other studies. In particular, these molecular emissions may trace the molecular components in torus/disk structures in the nucleus.

Acknowledgements

The European VLBI Network is a joint facility of European, Chinese, South African and other radio astronomy institutes funded by their national research councils. The Westerbork Synthesis Radio Telescope is operated by ASTRON (Netherlands Foundation for Research in Astronomy) with support from The Netherlands Foundation for Scientific Research (NWO).

References

Aalto, S., Polatidis, A.G., Hüttemeister, S. and Curran, S.J.: 2002, *A&A* **381**, 783.
Araya, E., Baan, W.A. and Hofner, P.: 2004, *ApJ* in press.
Baan, W.A.: 1989, *ApJ* **338**, 804.
Baan, W.A. and Haschick, A.D.: 1995, *ApJ* **454**, 745.
Baan, W.A., Haschick, A.D. and Henkel, C.: 1989, *ApJ* **346**, 680–689.
Baan, W.A., Rhoads, J., Fisher, K., Altschuler, D.R. and Haschick, A.D.: 1992, *ApJ* **396**, L99.
Baan, W.A., Wood, P.A.D. and Haschick, A.D.: 1982, *ApJ Lett.* **260**, L49–L52.
van Breugel, W.J.M., Stanford, S.A., Spinrad, H., Stern, D. and Graham, J.R.: 1998, *ApJ* **502**, 614.
Carilli, C.L. and Taylor, G.B.: 2000, *ApJ* **352**, 95.
Darling, J. and Giovanelli, R.: 2001, *AJ* **121**, 1278.
Genzel, R. et al.: 1998, *ApJ* **498**, 579.
Hagiwara, Y., Klöckner, H.-R. and Baan, W.A.: 2004, *A&A* in press.
Heisler, C.A., Lumsden, S.L. and Bailey, J.A.: 1997, *Nature* **385**, 700.
Henkel, C. and Wilson, T.L.: 1990, *A&A* **229**, 431.
Hüttemeister, S., Henkel, C., Mauersberger, R., Brouillet, N., Wiklind, T. and Millar, T.J.: 1995, *A&A* **295**, 571.
Klöckner, H.-R.: 2004, PhD Thesis; Rijksuniversiteit Groningen.
Klöckner, H.-R. and Baan, W.A.: 2004, *A&A* **419**, 887–896.
Klöckner, H.R., Baan, W.A. and Garrett, M.A.: 2003, *Nature* **421**, 821.
Lonsdale, C.J., Lonsdale, C.J., Diamond, P.J. and Smith, H.E.: 1998, *ApJ Lett.* **493**, L13.
Pedlar, A., Muxlow, T.W.B., Smith, R., Thrall, H., Beswick, R.J., Argo, M.K., Aalto, S., Booth, R.S. and Wills, K.A.: 2004, 'OH molecules & masers in M82'. in: S. Aalto, S. Hüttemeister, A. Pedlar (eds.), ASP Conference Series, The neutral interstellar medium of starburst galaxies, **320**, in press.
Pihlström, Y., Baan, W.A., Darling, J. and Klöckner, H.R.: 2004, *ApJ*, submitted.
Pihlström, Y., Conway, J.E., Booth, R.S., Diamond, P.J. and Polatidis, A.G.: 2001, *A&A* **377**, 413.
Richards, A.M.S., Cohen, R.J., Cole, G.H. et al.: 2001, *IAU Symposium* **205**, 212.
Rovilos, E., Diamond, P.J., Lonsdale, C.J., Lonsdale, C.J. and Smith, H.E.: 2003, *MNRAS* **342**, 373.
Scoville, N.Z. et al: 1998, *ApJL* **492**, 107.
Smith, H.E., Lonsdale, C.J., Lonsdale, C.J., Diamond, P.J.: 1998, *ApJL* **493**, L17.
Schilke, P. et al.: 1992, *A&A* 256, 595.
Solomon, P.M., Downes. D. and Radford, S.J.E.: 1992, *ApJ* **387**, L55.
Ulvestad, J.S., Wrobel, J.M. and Carilli, C.L.: 1999, *ApJ* **516**, 127.
Vernet, J., et al.: 2001, *A&A* **366**, 7.
Xia, X.Y., Xue, S.J., Mao, S., Boller, Th., Deng, Z.G. and Wu, H.: 2002, *ApJ* **564**, 196.

UNDERSTANDING EXTRAGALACTIC HYDROXYL

HANS-RAINER KLÖCKNER[1,2] and WILLEM A. BAAN[2]

[1]*Kapteyn Institute, University of Groningen, P.O. Box 800, Groningen, The Netherlands;*
E-mail: hrkloeck@astro.rug.nl

[2]*NFRA/ASTRON, P.O. Box 2, Dwingeloo, The Netherlands*

(Received 16 April 2004; accepted 15 June 2004)

Abstract. Hydroxyl (OH) is one of the few molecules in space showing both maser emission and absorption. In the Milky Way, hydroxyl has been found in various environments such as stellar envelopes, star-forming regions, and HII regions, tracing the physical states of the interstellar medium. Extragalactic OH has been found to provide essential information about the dusty and obscuring material within the inner kiloparsec region of active galaxies. At angular resolution of a few tens of parsecs, OH shows a rather complicated spatial distribution and the interpretation of the circumnuclear environment is rather difficult. Based on results obtained from low- and high-resolution observations, the diagnostic tools of OH will be reviewed and the strategies to further investigate extragalactic OH will be discussed.

Keywords: extragalactic molecule, hydroxyl, maser emission

1. Introduction

Hydroxyl (OH) and water vapor (H_2O) are the *classical* masering molecules for which most research has been done in the galaxy. OH emission can be found in star-forming regions, planetary nebulae, circumstellar disks, and stellar envelopes. In such environments, the various transitions of OH provide a perfect tool to study the physical condition of the surrounding medium at scale sizes ranging from astronomical units to hundreds of parsecs. The Galactic ground-state transitions with an extragalactic counterpart are found to be signposts of star formation or of violent ISM. The OH main lines (1667 and 1665 MHz) are seen toward HII regions and trace the molecular environment (Lockett et al., 1999). The 1720-MHz transition is confined within shells around SNR and probes the shock material (Wardle, 1999).

In the late 1970s, the first extragalactic OH main-line maser emission showed slightly higher luminosities than galactic ones and were associated with HII or star-forming regions displaced from the galactic centers (e.g. M 82, Nguyen-Q-Rieu et al., 1976). These extragalactic OH emission lines show distinctly different properties than the galactic lines such that the 1667 MHz main-line is found to be enhanced compared to the 1665 MHz and is unpolarized (Killeen et al., 1996). Soon after, a new sort of extragalactic main-line maser emission was discovered with

Astrophysics and Space Science **295:** 277–283, 2004.

isotropic luminosities of six or more orders of magnitudes higher than the galactic sources. Based on the extreme luminosity and the exceptional line width, it has been speculated that this megamaser emission traces the circumnuclear environment and exposes the nuclear properties of the host galaxies (Baan et al., 1982).

2. Indirect Diagnostics of OH Emission

Extragalactic OH Megamaser (OHMM) emission can be found among (ultra-) luminous FIR galaxies (ULIRG), which are generally characterized by morphological peculiarities, interactions with neighbors, highly ionized species showing starburst and AGN-type nuclear phenomena. Here we concentrate on the maser emission and what it tells us about the ISM.

The OH Megamaser emission spectrum shows a continuous (smooth) emission profile over a velocity range of up to thousands of kilometers per second and sometimes shows overlapping OH main-line features. While the H_2O Megamaser spectra are made up of predominantly narrow line components, the smooth OH spectrum indicate that it is made up of a superposition of broader line emission features. The number of contributing emission components can be estimated by comparing the observed line width of a distinct line feature or the total velocity coverage of the emission profile with the typical line width of the non-nuclear OH emission feature of 10 km s^{-1} seen in M 82 (Nguyen-Q-Rieu et al., 1976). The number can also be determined *via* the line width of a single emission line feature that is produced in a stationary layer of gas at a certain kinematic temperature. The line width at full width halve maximum (FWHM) (km s^{-1}) is

$$\Delta V_{\rm FWHM} = 2\sqrt{\frac{kT}{m_{\rm OH}}\ln(2)} \simeq 0.216 \times \sqrt{T},$$

where $m_{\rm OH}$ is the mass of the hydroxyl molecule ($m_{\rm OH} = 0.984$ amu $= 1.66 \times 10^{-27}$ kg). For a kinetic temperature of about $T = 61$ K, the expected line width is of the order of 1.7 km s^{-1}. For a typical line width of around 1257 km s^{-1} of OH emission seen in Mrk 273 (Klöckner, 2004; Klöckner and Baan, 2004) these basic estimates provide lower limits of about 125 individual OH emission regions or 740 individual clouds. Such estimates provide a good starting point for modeling the OH emission by individual clouds spread throughout the nuclear region, which will be reviewed in the next paragraph.

Another indirect diagnostic tool is based on the infrared emission seen in OH Megamaser galaxies. For the sampled galaxies, the OH emission is correlated with the infrared emission, which indicates that the energy source responsible for the excitation of the OH molecules, and therefore the maser emission, is the intense far-infrared radiation (Baan, 1985). In order to investigate the nature of the OH emission process itself, the conversion factor of the infrared photons into OH maser photons can be evaluated. Considering the spectral width of the OH emission and applying this to the infrared emission, a lower limit of the OH pumping efficiency can be estimated using the following ratio:

$$P_{\rm OH} = \frac{L_{\rm OH} \times r_{\rm FIR}}{L_{\rm FIR} \times r_{\rm OH}},$$

where $L_{\rm OH}$ and $L_{\rm FIR}$ are the luminosities and $r_{\rm OH}$ and $r_{\rm FIR}$ are the spectral ranges of the emission. For this ratio, an averaged conversion factor of about $\sim$0.2% has been found for OH Megamaser galaxies (Klöckner, 2004). Such low pump efficiency indicates an unsaturated maser process for the overall emission. Because maser emission would be saturated if every available pumping event (in this case every infrared photon) produces a maser photon with an efficiency that depends only on the details of the pumping environment. This efficiency provides a further determination at which amplification or gain the saturation effects needs to be taken into account. A rough estimate of this amplification can be determined from the ratio γ of the maser intensity at with saturation takes place. At this stage, the ratio is about unity:

$$1 \simeq \gamma = \frac{I_{\rm sat}}{I_{\rm unsat}} \sim \frac{P_{\rm OH}\Gamma}{A_{2-2+}}$$

where $I_{\rm sat}$ is the intensity at saturation of the maser, $I_{\rm unsat}$ the unsaturated maser intensity, A_{2-2+} the Einstein coefficient of the 1667 MHz transition, $P_{\rm OH}$ the efficiency as defined earlier, and Γ the loss rate (Elitzur, 1992). The loss rate Γ is generally of the same order as the collision rate ($\Gamma = 10^{-10}$ N s^{-1}), which can be found for the various OH transitions in the literature (Offer and van Dishoeck, 1992; Offer et al., 1994). Assuming a general OH density N of 10^6 cm^{-1} and the averaged value of the efficiency, saturation will take place when the gain exceeds $\ln(\gamma) > 10.6$. Such amplification factors have not been detected in extragalactic sources and together with the low efficiency, this clearly indicates that extragalactic OH emission emerges from an unsaturated maser process. This altogether assures that for modeling OH emission in extragalactic sources, some simplified conditions of the pumping environment may be assumed.

3. Understanding Extragalactic OH Emission

Since the first high-resolution observations of OH Megamaser sources it has become clear that some of the line and continuum emission characteristics can be studied in great detail within the nuclear region of active galaxies (Diamond et al., 1999). But so far little has been done to interpret the detected OH main-line emission at such resolutions and the theoretical modeling has mostly been concentrated on galactic sources. A complicating factor in such theoretical investigations is that the OH maser emission shows components with varying degrees of compactness at a resolution of a few parsecs. Furthermore, some of these observations resolve almost ~90% of the total continuum and some ~50% of the OH flux measured with a single-dish or a low-resolution interferometer array. Examples of this effect are found in Arp 220, IRAS 17208-0014 and IRAS 10039-3338 observed with global VLBI/VLBA (Lonsdale et al., 1998; Diamond et al., 1999; Rovilos et al., 2002) or in the nearby sources IC 694 observed with the EVN (Klöckner and Baan, 2002). In addition, the interpretation of the OH pumping may also be complicated by an observed 1720 MHz satellite line in some of the OHMM source (Baan et al., 1989). This satellite line has been seen in association with the blue-shifted wings of the 1667 MHz main-line features, suggesting that the emission at such velocities traces a physically different environment. However, observations at the highest angular resolution do not reveal this blue-shifted components and their contribution is weak compared to the strong main-line emission. All this makes it difficult to interpret and understand these unusual OH emission components, but it clearly suggests that the OH emission occurs at various scales between a few tens of parsecs and up to hundreds of parsecs.

In the previous section, it has been indicated that the low efficiency and the nature of the maser emission reduce the need for a sophisticated model to describe the OH radiation field. Here a basic approach has been proposed to describe the observed OH emission spectrum by using a superposition of individual clouds with an emissivity of the order unity. A similar approach has been used to describe the OH emission in the source III Zw35 (see R. Parra et al., 2004 this issue). First results are shown that simulate a representative emission pattern seen in OHMM. To this end a three-dimensional disk and geometrical torus structure have been modeled to solve for the line-of-sight maser amplification. In these geometries, the amplification is calculated on the basis of randomly distributed maser clouds with optical depths of unity within a Keplerian disk of about 450 pc in extent. The resulting emission spectrum has been determined by a Monte-Carlo approach by integrating over large number of synthesized spectra (see Figure 1, left panel). For a prototype emission pattern of a triple line, with an enhanced central line feature, the individual geometries need to be seen edge-on within a range of 15deg and a half obscuration angle between 50 and 30°. To produce the smooth emission line spectrum that is common for all OH Megamaser galaxies, the individual OH clouds need to be extended and to exhibit internal velocity dispersion or turbulence. This model suggests a minimal extent of maser clouds of the order of >10 pc (Klöckner, 2004).

Such modeling provides good constrains on how the emission structure is built up, but it does not give a good indication of how the emission is generated and how much the 1665 MHz main-line emission contributes to the synthesized spectrum. In order to understand this contribution, the main-line ratio needs to be investigated. An overview of the pumping scheme of OH main-line emission can be found in the studies of circumstellar maser emission (Collison and Nedoluha, 1993; Bujarabal

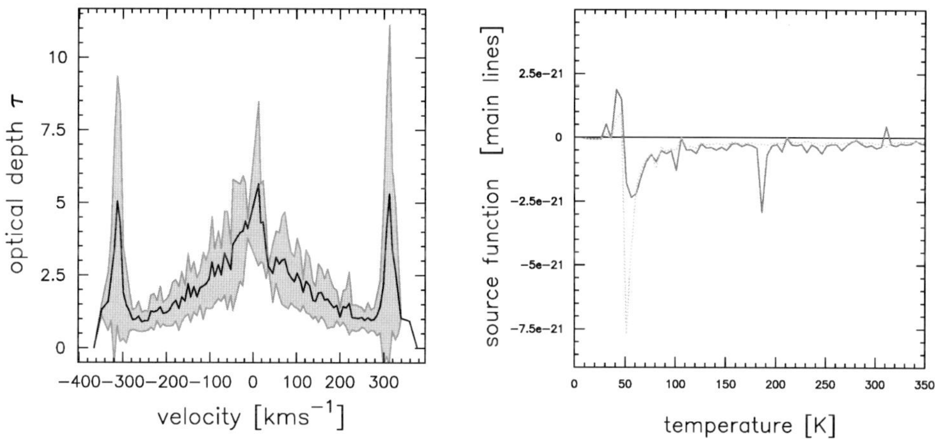

Figure 1. Left panel: A synthesized emission spectrum for 1300 randomly distributed clouds of ∼28 pc in size within a torus geometry. The model parameters for the torus have been set to a maximal extent of 450 pc, a central mass of $5 \times 10^9 M_\odot$ with a Keplerian velocity field and an obscuration angle of 104°. *Right panel*: Source functions of the 1667 MHz (solid line) and 1665 MHz (dotted line) main-lines transitions versus temperature of a diluted and modified black body (dilution factor is 0.1%). Note that the source functions are normalized to unity.

et al., 1980; Elitzur, 1978) or in a more general investigations of OH environments (Pavlakis and Kylafis, 1996; Randell et al., 1995; Cesaroni and Walmsley, 1991; Bujarrabal et al., 1980). In these models, the pumping scheme has been considered in relation to collisions, local line overlap, non-local line overlap, and external infrared radiation fields. And the steady-rate equation has been solved in terms of an escape probability formalism and a modified Sobolev approximation. Here a more simplified model has been used to understand the main-line ratios observed at highest resolution in OH Megamaser sources.

Therefore, the pumping scheme has been investigated based on an external infrared radiation field and by neglecting the effects of local and non-local line overlaps and collisions. Such simplifying assumption may be justified because of the success of modeling the integrated velocity pattern by using individual clouds: (i) A low number of individual clouds assure sufficient velocity offsets such that local and non-local line overlap in the radio- and the infrared bands will have little influence to the level populations. (ii) The individual clouds represent an environment where only a few collision partners of the OH are present and if the density of the collision partners is of the order of $<10^4$ cm^{-3} the effects of collisions can be entirely neglected (Randell et al., 1995). In case of higher densities ($>10^6$ cm^{-3}) line features of the satellite transitions are expected, whereas the main lines show an almost thermal main-line ratio (van Langevelde et al., 1995). Here main-line maser action is produced for pure radiative pumping by a modified black body emission, by solving the steady-state equations and by using an escape probability of unity for all transitions. With these assumptions, the level population distributions are independent of the column density, which best describes the situation of unsaturated OH Megamaser. The steady-state-rate equation can than be written as:

$$0 = \sum_{j>k} g_j A_{jk}[n_j - F(\nu_{jk})(n_k - n_j)] - \sum_{j<k} g_k A_{kj}[n_k - F(\nu_{kj})(n_j - n_k)],$$

where A_{jk} are the Einstein coefficients, n_j and n_k designate the individual levels, g_j and g_k designate the weight of the levels, and F describes the radiation field resulting from a superposition of a 2.7 K black body and a diluted (0.1%) and modified black body emission (see description in Elitzur, 1978 and for the modified black body Klass et al., 2001).

Preliminary results are shown in Figure 1, right panel, where the source function of the main-line emission indicates maser emission at temperatures greater than 48 K. Two ranges are found with a width of around 15 K where the OH main-line emissions show significant differences in their source functions. Between 48 and 61 K, both source functions varies, whereas between 181 and 195 K only the 1667 MHz source function shows variations. In the first range, an enhanced 1667 MHz line can be seen with main-line ratios of about ~4. This temperature of around 60 K is indeed found in OH Megamaser sources and give a good constraint for the assumptions made here Klass et al. (2001).

4. Conclusions

An overview has been given of the characteristics of the OH Megamaser emission and the difficulties of interpreting such emission. It has been shown that basic modeling of optically thin masering clouds provides a good representation of the velocity pattern and the structure seen in OH Megamaser sources. Furthermore, the OH emission may indeed be a superposition of individual clouds spread within the circumnuclear region due to an unsaturated maser emission process. The pumping agent of the OH main-lines in megamasers can be described by a diluted black body emission in the infrared.

For a more elaborated investigation of circumnuclear OH, further high-resolution observations are needed to cover the different scale sizes from parsecs to hundreds of parsecs. Based on such observations, an even more realistic picture may be drawn by involving radiative transfer modeling of the maser emission in combination of the modeling with dusty circumnuclear environment.

References

Baan, W.A.: 1985, *Nature* **315**, 26.
Baan, W.A., Haschick, A.D. and Henkel, C.: 1989, *ApJ* **346**, 680.
Baan, W.A., Wood, P.A.D. and Haschick, A.D.: 1982, *ApJ Lett.* **260**, L49.

Bujarrabal, V., Guibert, J., Rieu, N.-Q., Destombes, J.L., Marliere-Demuynck, C. and Omont, A.: 1980, *A&A* **81**, 1.
Cesaroni, R. and Walmsley, C.M.: 1991, *A&A* **241**, 537.
Collison, A.J. and Nedoluha, G.: 1993, *ApJ* **413**, 735.
Diamond, P.J., Lonsdale, C.J. and Smith, H.E.: 1999, *ApJ* **511**, 178.
Elitzur, M.: 1978, *A&A* **62**, 305.
Elitzur, M.: 1992, Astronomical Masers, Astrophysics and Space Science Library, Kluwer Academic Publishers, Vol. 170, 365 p.
Killeen, N.E.B., Staveley-Smith, L., Wilson, W.E. and Sault, R.J.: 1996, *MNRAS* **280**, 1143.
Klaas, U., Haas, M., Müller, S.A.H., Chini, R., Schulz, B., Coulson, I., Hippelein, H., Wilke, K., Albrecht, M. and Lemke, D.: 2001, *A&A* **379**, 823.
Klöckner, H.-R.: 2004, Ph.D. Thesis, Rijksuniversiteit Groningen.
Klöckner, H.-R. and Baan, W.A.: 2002, in: V. Migenes and M.J. Reid (eds.), IAU Symposium 206: Cosmic Masers: From protostars to blackholes, 2001, Mangaratiba, Rio De Janeiro, Brazil, Astronomical Society of the Pacific, San Francisco, pp. 430.
Klöckner, H.-R. and Baan, W.A.: 2004, *A&A* **419**, 887.
Lockett, P., Gauthier, E. and Elitzur, M.: 1999, *ApJ* **511**, 235.
Lonsdale, C.J., Diamond, P.J. and Smith, H.E.: 1998, *ApJ Lett.* **493**, L13.
Nguyen-Q-Rieu, U., Mebold, A., Winnberg, J., Guibert, and Booth, R.: 1976, *A&A* **52**, 467.
Offer, A.R. and van Dishoeck, E.F.: 1992, *MNRAS* **257**, 377.
Offer, A.R., van Hemert, M.C. and van Dishoeck, E.F.: 1994, *J. Chem. Phys.* **100**, 362.
Pavlakis, K.G. and Kylafis, N.D.: 1996, *ApJ* **467**, 300.
Randell, J., Field, D., Jones, K.N. Yates, J.A. and Gray, M.D.: 1995, *A&A* **300**, 659.
Rovilos, E., Diamond, P.J., Lonsdale, C.J. and Smith, H.E.: 2002, New Developments in VLBI Science and Technology, 179 pp.
van Langevelde, H.J., van Dishoeck, E.F., Sevenster, M.N. and Israel, F.P.: 1995, *ApJ Lett.* **448**, L123.
Wardle, M.: 1999, *ApJ Lett.* **525**, L101.

THE SUB-PC SCALE ACCRETION DISK OF NGC 4258

E.M.L. HUMPHREYS, A.L. ARGON, L.J. GREENHILL, M.J. REID
and J.M. MORAN

Harvard-Smithsonian Center for Astrophysics, 60 Garden Street, Cambridge, MA, USA;
E-mail: ehumphreys@cfa.harvard.edu

(Received 16 April 2004; accepted 15 June 2004)

Abstract. Water megamasers have been found to trace parsec/sub-parsec, circumnuclear accretion disks in several AGN (e.g., Circinus, NGC 1068 & NGC 4258). High-spatial (0.5 mas) and velocity resolution (0.2 km s^{-1}) VLBA imaging of the disks reveals thin, warped 'pannekoeken (pancake)'-style structures as opposed to thick tori in the inner regions of the central engines (40 000 R_{sch}). In this contribution, I will describe some current investigations into the dynamical and physical attributes of the water maser disk in NGC 4258, as revealed by VLBA, VLA and Effelsberg monitoring over 8 years.

Keywords: NGC 4258, water masers, accretion disks, AGN

1. Introduction

NGC 4258 is a nearby (∼7.2 Mpc) Seyfert II/LINER that has been the subject of extensive study over the past decade due to the discovery of a sub-parsec scale accretion disk traced out by 22 GHz water maser emission. Due to the relatively simple edge-on geometry of the source, and the almost perfect Keplerian rotation of the maser disk (to ∼1%) (Miyoshi et al., 1995, Herrnstein et al., 1999), NGC 4258 currently provides a unique opportunity for obtaining a detailed understanding of the physics of accretion disks and AGN.

2. Modeling the Maser Disk

We imaged NGC 4258 at 23 epochs over 6 years. In order to construct the 3D disk model, we employ a nine-parameter χ^2 fitting routine that fits for the sky positions and Doppler velocities of maser emission from VLBA epochs. The current best-fitting disk model is shown in Figure 1, in which the VLBA data from a subset of 12 epochs is included in the current analysis. Table I shows that the best fit to date indicates that *the maser disk is significantly warped* (Herrnstein et al., 1999, 2004; Humphreys et al., 2003).

Astrophysics and Space Science **295:** 285–289, 2004.

TABLE I

Current best-fit parameters for maser disk, where the label definitions are: systemic, galactic velocity (v_{sys}), dynamical disk center (x_0, y_0), black hole mass/distance ($\mathcal{M} = M/D$), position angle warping ($\alpha(\theta_r) = \alpha_0 + \alpha_1\theta_r + \alpha_2\theta_r^2$; α measured East of North) and inclination angle warping ($i(\theta_r) = i_0 + i_1\theta_r$) where θ_r is the radial distance in the disk in angular units

v_{sys} (km s^{-1})	x_0 (mas)	y_0 (mas)	M/D ($10^7 M_\odot$ Mpc^{-1}	α_0 (deg)	α_1 (deg mas^{-1})	α_2 (deg mas^{-2})	i_0 (deg)	i_1 (deg mas^{-1})
474.1	−0.132	0.548	0.526	65.3	2.3	−0.23	71.6	2.4

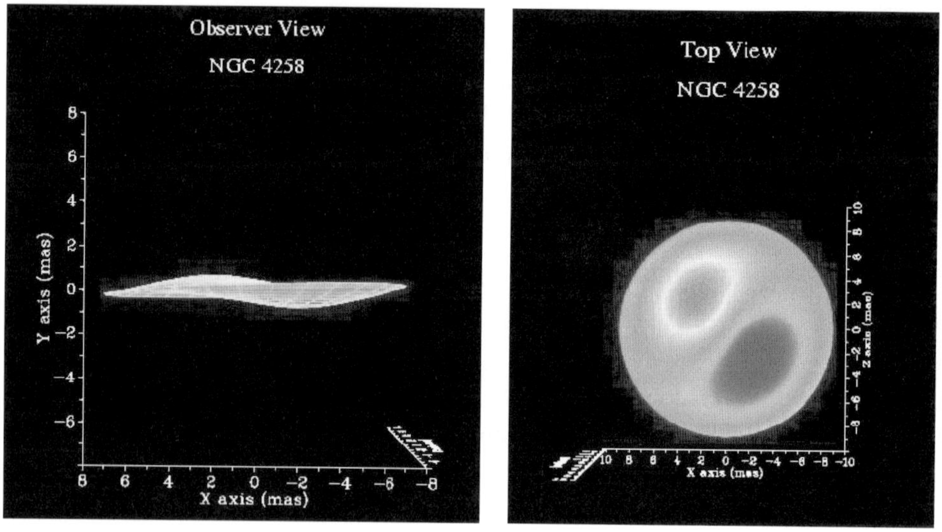

Figure 1. 3D model of NGC 4258 accretion disk. (a) The observer's view (observer located at (0,0, large negative z)); (b) the view looking down on the maser disk from above, such that the rotation sense of the disk appears clockwise.

3. Current Investigations

3.1. Why is the Disk Warped?

Potential causes of warping in the NGC 4258 maser disk are discussed by Moran et al. (1999), and include: (i) A binary companion orbiting outside the maser disk and of comparable mass to that of the disk ($<10^6$ $M_\odot$) (Papaloizou et al., 1998). However, Papaloizou et al. (1998) find that a *trailing* warp twist would result for this scenario, rather than the leading twist that we find here, see Figure 1b. (ii) The warp could be due to the Lense Thirring (LT) effect, i.e., drag on material orbiting the black hole that occurs when the material's orbit is inclined with respect to the equatorial plane of the rotating black hole. For the assumption of a maximally rotating black hole and a Shakura–Sunyaev thin disk (Shakura and Sunyaev, 1976), the change in the longitude of nodes caused by the LT precession is ~6° (Trotter, private communication) across the high-velocity features in the maser disk. This is of the same order as the change in disk inclination angle across the high-velocity features (~9°), and we have not ruled this out as the origin of the disk warp. (iii) Finally, an already slightly warped disk can become more warped due to torques on the disk caused by radiation pressure from the central source (Maloney et al., 1996). In Circinus, a more luminous source, Greenhill et al. (2003) find a greater degree of disk warping further from the black hole, which would be consistent with a scenario in which radiation pressure determines the nature of warping in AGN accretion disks.

3.2. Measurement of the Jerk

We have a current observing program aimed at measuring the jerk, i.e. the time derivative of acceleration. This quantity is important since the accurate geometric distance measurement to NGC 4258 (currently 5% uncertainty) is based on assuming ballistic motion of maser features in the disk. The measurement of the jerk would provide an *independent* piece of evidence in support of this interpretation. Moreover, the measurement of the jerk would provide a new and independent means of obtaining a geometric distance to the source. See Figure 2 for the expected value of the jerk. Figure 3 shows that we are able to obtain comparable model fits to the bulk velocity trends in our data by including (i) a linear acceleration only (Figure 3a), or (ii) the inclusion of the higher order jerk term in the model (Figure 3b). Since

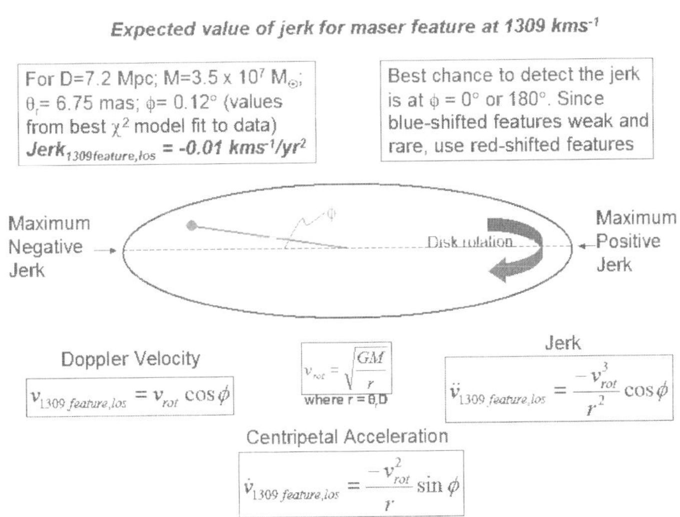

Figure 2. Estimated value of the jerk for a red-shifted spectral feature near the midline.

OH MEGAMASERS: CIRCUMNUCLEAR GAS IN STARBURSTS

Y.M. PIHLSTRÖM

National Radio Astronomy Observatory, New Mexico, USA; E-mail: ypihlstr@nrao.edu

(Received 16 April 2004; accepted 15 June 2004)

Abstract. OH megamaser emission is known to be associated with the most central 100 parsecs of ultra-luminous infrared galaxies (ULIRGs). The masers are probing dense concentrations of gas, which presumably have accumulated during the merger event forming the ULIRG. Here we summarize the results from primarily high resolution observations of the 18 cm OH lines in a few OH megamaser galaxies. The maser emission is commonly distributed in disks or tori, but there is also some evidence for other components such as outflows. The appearance of the observed emission strongly depends on the observing resolution, and new theoretical modeling has made important steps toward understanding the observed emission. The radio continuum in OH megamaser galaxies is generally starburst related, and detailed studies of the continuum could perhaps be used to date the merger.

Keywords: maser, OH, starburst, ULIRG

1. Introduction

The extremely luminous OH maser emission detected in extragalactic sources has been shown to be associated with the most central regions of ultraluminous infrared galaxies (ULIRGs). ULIRGs are known to be merging systems (e.g. Clements et al., 1996), and as a result they display very high FIR to optical ratios and $L_{\rm FIR} > 10^{11} L_{\odot}$. ULIRGs can be linked to a number of fundamental research areas such as globular cluster formation, the formation of elliptical galaxies, the formation of quasars and the high-z proto-galaxies detected by SCUBA and MAMBO. For instance, studies of ULIRGs allow investigation of the possible evolutionary relationship between AGN and ULIRGs (the starburst-AGN connection), since both AGN and starburst activity are suggested to be the result of mergers. Other examples of specific research topics include galaxy evolution scenarios, and how star formation and AGN activity might affect the host galaxy and the intergalactic medium.

In the local universe, ULIRGs dominate the extragalactic source population with luminosities $>10^{11} L_{\odot}$, which makes them obvious targets when investigating all of the research areas mentioned above. However, the cores of the ULIRGs are heavily dust-obscured, which complicates observations of these objects in the optical, UV and NIR regimes. As the radio emission is not much affected by dust, observing the 18 cm OH maser lines with very long baseline interferometry (VLBI) can pinpoint the innermost dynamics and gas distribution in the ULIRGs.

a rotating ring structure. Morphologically the diffuse and compact masers appear to belong to the same structure, and it is then straightforward to understand the compact masers as positions on the sky where the path length of velocity coherent gas is maximised (Pihlström et al., 2001).

Parra et al. (2004; these proceedings) present new Monte-Carlo modeling of the masers in III Zw 35. Excitingly, these results show that it is possible to reconstruct the observed maser emission at every resolution using *a single gas phase and a clumpy medium.* Hence, it may not be necessary to use a two-phase gas to explain the high maser gains (see also Conway et al., 2004; these proceedings).

3. The Radio Continuum

Perhaps equally important to the OH emission in refining the maser model, is the radio continuum. Firstly, the radio emission will provide the seed photons which stimulate the maser emission. Secondly, mapping the radio emission could provide a way to trace the star forming regions which presumably are responsible for the pumping.

In Figure 1 we show the results from 1.6 GHz continuum imaging at two different resolutions of III Zw 35. The lower resolution image shows a smooth distribution, while at higher resolution there are a few compact clumps detectable, holding only a minor part of the total flux density (Pihlström et al., 2001). Similar behavior with a small part of the total continuum flux density residing in a

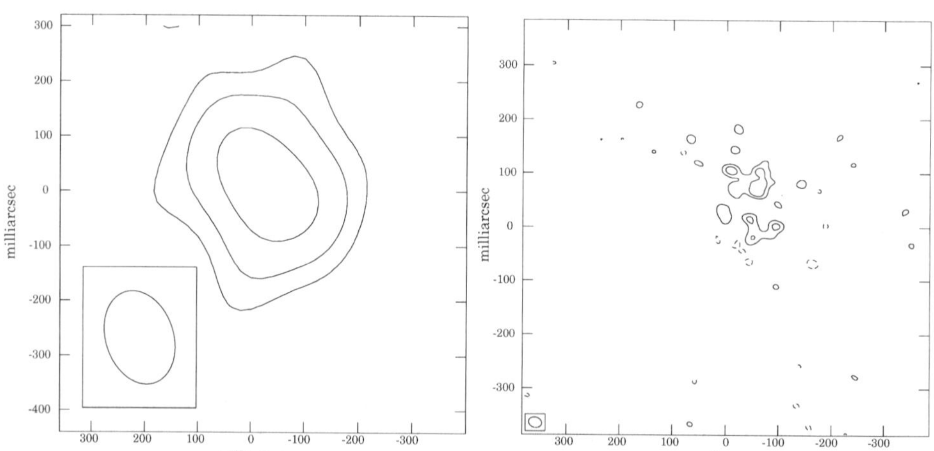

Figure 1. The distribution of the radio continuum emission in the OH megamaser galaxy III Zw 35. *Left:* MERLIN only data at a resolution of 174 × 127 mas. *Right:* EVN-only naturally weighted map, resolution 25 × 19 mas. While the diffuse emission is consistent with being the synchrotron emission resulting from a population of old supernova remnants, the clump like emission observed at higher resolution could either be individual radio supernovae, or clumps of supernovae and/or supernova remnants.

number of very compact spots has also been observed in for instance Arp 220 and IRAS 17208-0014 (Smith et al., 1998; Momjian et al., 2003). Even though there are examples of OH megamaser galaxies harboring powerful radio sources (e.g. Mrk 231), from these observations it seems that most OH megamaser hosts have no compact AGN radio component. Instead, there appears to be two different continuum components, one diffuse and one that consists of a few compact clumps. In order to address the question of what the nature of the radio continuum emission is, we consider below the diffuse and the compact component in turn using III Zw 35 as an example.

3.1. THE DIFFUSE RADIO CONTINUUM EMISSION

As III Zw 35 follows the radio-IR correlation, Pihlström et al. (2001) used the FIR flux to estimate the star formation rate to 19 $M_{\odot}$ yr^{-1}. In turn, this yielded a supernova rate of 0.8 yr^{-1}. The authors found that the estimated, total emitted non-thermal flux from these supernova remnants closely agreed with the observed diffuse radio luminosity component. Hence, the diffuse radio emission is consistent with being due to the synchrotron emission from an old population of supernova remnants.

3.2. THE COMPACT RADIO CONTINUUM EMISSION

In Arp 220 the compact radio components most likely are individual radio supernovae (Lonsdale et al., 1998). Such radio supernovae would in our observations of III Zw 35 be detectable above the noise level for around 7 yr. Given the estimated supernova rate estimated we would thus expect to detect about 6 components. Even though this is in agreement with the number of clumps observed, this argument assumes that every supernovae must be of the very luminous type. An alternative explanation is therefore that the compact clumps could be *clumps of supernovae or supernova remnants*. Similar conclusions have been drawn for the clumpy radio continuum observed in IRAS 17208-0014 (Momjian et al., 2003). Strong support for clumpy starburst regions further comes for example from HST UV images (Meurer et al., 1995).

It is worth noting that if indeed the radio continuum results from two components, one of them tracing the old supernovae population, and the other a more recent starburst, then the ratio between these two components should vary with starburst age. Hence, measuring this ratio could provide a possibility to date the merger.

4. Summarizing Remarks

Observations of the 18 cm OH megamaser provide an unobscured view of the most central regions of merging galaxies. In particular, VLBI techniques can be used to probe the dynamics and structure of nuclear gas on scales of a few parsecs. The small sample of OH megamasers studied by VLBI so far clearly suggests that the OH maser arises in disks or tori. Current VLBI arrays do not yet have sufficient sensitivity and wide enough range of baselines to completely map a larger sample of mJy-level OH maser emission lines with sufficiently high signal-to-noise ratios. In the future, studies of the OH megamaser lines at 1667 and 1665 MHz will thus benefit from the EVLA upgrade, providing both increased sensitivity as well as bridging VLA and VLBI baselines. A significant improvement in sensitivity compared to current telescope arrays is also expected from the square kilometer array (SKA), which will allow detailed and extremely sensitive studies of OH megamasers.

Many questions still remain to be answered in order to better understand the physics of the OH megamasers. For instance, does OH megamaser emission occur in several physically different phases (Section 2)? Theoretical modeling that closely follows observations at all angular scales will obviously be particularly important to solve this issue. A related question which could help to constrain the modeling is whether the 1665 MHz masers are occurring at a different physical region than the 1667 MHz masers. The 1665 MHz masers are weaker than the 1667 MHz lines, and have mostly remained undetected in the VLBI experiments, implying that the 1665 MHz masers either are resolved out or too weak to be detected.

References

Baan, W.A.: 1985, *Nature* **315**, 26.

Baan, W.A., Haschick, A.D. and Henkel, C.: 1989, *ApJ* **346**, 680.

Briggs, F.H.: 1998, *A&A* **336**, 815.

Clements, D.L., Sutherland, W.J., McMahon, R.G. and Saunders, W.: 1996, *MNRAS* **279**, 477.

Condon, J.J., Huang, Z.-P., Yin, Q.F. and Thuan, T.X.: 1991, *ApJ* **378**, 65.

Darling, J. and Giovanelli, R.: 2002a, *AJ* **124**, 100.

Darling, J. and Giovanelli, R.: 2002b, *ApJ* **572**, 810.

Diamond, P.J., Lonsdale, C.J., Lonsdale, C.J. and Smith, H.E.: 1999, *ApJ* **511**, 178.

García-Burillo, S., Martín-Pintado, J., Fuente, A. and Neri, R.: 2000, *A&A* **355**, 499.

Klöckner, H.-R., Baan, W.A. and Garrett, M.A.: 2003, *Nature* **421**, 821.

Lonsdale, C.J., Lonsdale, C.J., Diamond, P.J. and Smith, H.J.: 1998, *ApJL* **493**, L13.

Meurer, G.R., Heckman, T.M., Leitherer, C., Kinney, A., Robert, C. and Garnett, D.R.: 1995, *AJ* **110**, 2665.

Momjian, E., Romney, J.D., Carilli, C.L., Troland, T.H. and Taylor, G.B.: 2003, *ApJ* **587**, 160.

Montgomery, A.S. and Cohen, R.J.: 1992, *MNRAS* **254**, 23.

Norris, R.P., Baan, W.A., Haschick, A.D., Diamond, P.J. and Booth, R.S.: 1985, *MNRAS* **213**, 821.

Pihlström, Y.M., Conway, J.E., Booth, R.S., Diamond, P.J. and Polatidis, A.G.: 2001, *A&A* **377**, 413.

Rovilos, E., Diamond, P.J., Lonsdale, C.J., Lonsdale, C.J. and Smith, H.E.: 2003, *MNRAS* **342**, 373.

Sakamoto, K., Scoville, N.Z., Yun, M.S., Crosas, M., Genzel, R. and Tacconi, L.J.: 1999, *ApJ* **514**, 68.

Smith, H.E., Lonsdale, C.J., Lonsdale, C.J. and Diamond, P.J.: 1998, *ApJL* **493**, L17.

CONTINUUM EVN AND MERLIN OBSERVATIONS OF ULTRA LUMINOUS INFRARED GALAXIES

A.G. POLATIDIS[1], J.E. CONWAY[2], Y.M. PIHLSTRÖM[3] and R. PARRA[2]

[1]*Max Planck Institut für Radioastronomie, Bonn, Germany; E-mail: apolatid@mpifr-bonn.mpg.de*
[2]*Onsala Space Observatory, Sweden; E-mails: jconway@oso.chalmers.se, rodrigo@oso.chalmers.se*
[3]*NRAO, USA; E-mail: ypihlstr@nrao.edu*

(Received 16 April 2004; accepted 15 June 2004)

Abstract. Radio imaging of ULIR galaxies is ideal to explore the connection between the starburst and the AGN phenomenon since it is unaffected by dust obscuration, and provides the required high angular resolution to distinguish between an AGN and starburst emission. We have made combined 18 cm radio continuum, EVN and MERLIN observations of 13 ULIRGs that have the parsec and deci-parsec scale resolution necessary to distinguish between an AGN and supernovae remnants at the centres of these galaxies, and assess the contribution of each to the total energy distribution. Images of three galaxies are presented here.

Keywords: galaxies-active, galaxies-starburst, galaxies-nuclei

1. Introduction

To date the best-studied sample at high resolution of ULIRG galaxies is the IRAS Bright Galaxy sample (hereafter the BGS sample). Condon et al. (1990, 1991) observed over 40 compact sources with the VLA from 1.6 to 8 GHz. Lonsdale et al. (1993, L93) made snapshot VLBI images of 31 sources from Condon et al. (1991); of these 21 were detected in VLBI baselines greater than 10 Mλ, prompting Londsale et al. to suggest that an AGN was responsible for the VLBI structure. However, Smith et al. (1998a), in full track observations of Arp 220 showed that the VLBI structure consists of radio supernovae or supernovae remnants. Following this, Smith et al. (1998b) modelled the L93 data fitting clustered luminous compact radio supernovae, showing that for most sources the VLBI structure could be explained by starburst activity while for only a few sources the presence of an AGN was invoked to account for the parsec scale structure.

We are engaged in a systematic survey of the complete sample of the 19 northern ($\delta > 0$) compact ULIRGs in the BGS sample which showed evidence for detections >1 mJy on at least one baseline longer than 10 Mλ. Of these 19 compact ULIRGs six (IIIZw 35, Mrk 231, Mrk 273, Arp 220, NGC 7469 and Mrk 331) were already well observed at the same resolution, hence we were left with 13 objects to observe.

Astrophysics and Space Science **295:** 297–302, 2004.

We have observed the sources with a combined MERLIN and EVN array at 18 cm so that we can separate the diffuse emission at MERLIN scales which can be due to thermal emission combined with synchrotron emission from electrons from supernovae diffused into the ISM, from the compact radio emission which can be due to an AGN or to synchrotron emission associated with single or clustered radio supernovae or supernovae remnants (SNRs).

2. Observations

Initially the five brighter sources were observed with the combined MERLIN and EVN array, in February 2002. The remaining eight sources were observed in May 2003. In this paper we present preliminary results from February 2002 observations.

The sources were observed with six telescopes of the EVN (Torun, Medicina, Onsala, Effelsberg, Westerbork, and the Jodrell Bank Lovell and Mark 2 telescopes alternating during the observations) with a sampling rate of 256 Mbits/s over a 32 MHz bandwidth. The MERLIN observations had a 15 GHz bandwidth in overlapping frequency range. Due to the relative weakness of the sources, the observations were made in phase-referencing mode with a cycle of 7 min on the source and 3 min on the nearby phase calibrator.

The observations were partially successful. Of particular importance was the failure of the Cambridge telescope which would have provided the common baseline (Lovell, Cambridge) to facilitate the cross-calibration of the two arrays, and the subsequent reduction of the MERLIN resolution to ~250 mas. Furthermore the loss of the Lovell telescope for 40% of the run, although it was replaced by the Jodrell Bank Mark 2 antenna, resulted in reduction of sensitivity of the array. Consequently the array consisted of the 6 MERLIN telescopes and 6 EVN antennae at the most for each source.

3. Results

The EVN data were correlated at the JIVE data processor and were then processed with the AIPS software suite. Initial amplitude calibration and fringe fitting was done with the EVN pipeline and was refined later.

3.1. UGC 05101

UGC 05101 at VLA scales shows a ~1.57 ×1.14 arcsec (1110 × 810 pc, at a distance $d = 158$ Mpc) disk-like emission elongated roughly east-west, dominated by a central bright component (seen clearly in Condon et al.'s (1991) 8.4 GHz image). At the MERLIN resolution of 270 mas (Figure 1) we detect a similar structure measuring 933 × 533 pc (roughly 1.3 × 0.75 arcsec). The flux density in the MERLIN image is consistent with that of Condon et al. (1990) and Thean et al. (2001).

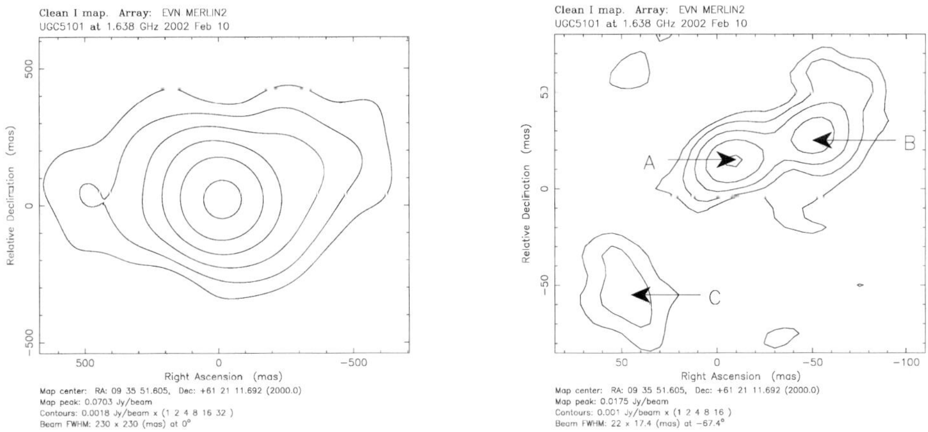

Figure 1. Left: The MERLIN image of UGC 05101 with a restoring beam of 230 mas. The peak is 70.3 mJy/bm and the contours are drawn at 1, 2, 4, 8, 16, 32 $\times$ 1.8 mJy/bm. Right: The EVN image of UGC 05101 restored with a beam of 22 $\times$ 17.4 mas. The peak is 17.5 mJy/bm and the contours are drawn at 1, 2, 4, 8, 16 $\times$ 1 mJy/bm.

In the highest resolution (15 pc) EVN image $\sim$53% of the flux density is recovered in an apparently S-shaped source extending roughly 128 $\times$140 parsec. Three major components (A–C), relatively compact (5–13 pc) are embedded in weak extended radio emission, visible in the lower resolution images that are not presented here. Component A is closer to the peak brightness of the MERLIN image. The brightness temperatures of components A and B are $\sim 10^7$ K while that of the more resolved component C is 7 $\times$ 10^6 K, while their luminosities range from 5 $\times$ 10^{39} to 10^{40} erg/s.

Lonsdale et al. (2003) with higher resolution 5 GHz global VLBI observations find a consistent nuclear structure, while they resolve A and B in two compact subcomponents each.

These compact components are $\sim$50 times more luminous than the brightest radio supernovae (RSN) in Arp 220. They are also resolved and their sizes are too large for single RSN. It appears that the nuclear structure of UGC 05101 can not be associated with RSN but is rather caused by an obscured AGN. This is consistent with Imanishi et al.'s (2003) detection of Fe Kα line emission in X-rays, though the X-ray spectrum was measured with an aperture of 8 arcsec (5.6 kpc) which covers the whole radio source and hence it is not easy to correlate the position of the X-ray source and the radio emission. Multi-frequency observations are needed to further explore this possibility.

3.2. NGC 6286

NGC 6286 is an edge-on spiral galaxy (at a distance d = 73 Mpc) in an interacting pair with NGC 6285 (1′.5, 30 kpc) classified as a LINER. Condon et al. (1990)

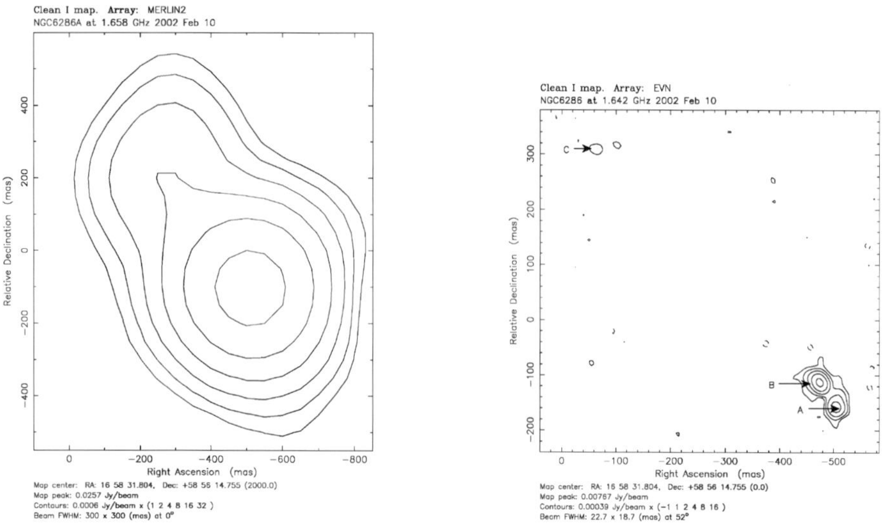

Figure 2. Left: The MERLIN image of NGC6286 with a restoring beam of 300 mas. The peak is 25.7 mJy/bm and the contours are drawn at 1, 2, 4, 8, 16, 32 $\times$ 0.6 mJy/bm. Right: The EVN image of NGC6286 restored with a beam of 22.7 $\times$ 18.77 mas. The peak is 7.6 mJy/bm and the contours are drawn at 1, 2, 4, 8, 16 $\times$ 0.39 mJy/bm.

interpret the 1.6 GHz VLA image, showing a 20″ (6.8 kpc) resolved source elongated along the optical axis as starburst induced.

Our MERLIN image (Figure 2) recovers 37 mJy (~25% of the VLA flux density) and consists of two components separated by 450 mas (150 pc) at the same orientation as the VLA image. At the 25 mas resolution of the EVN image, the brighter MERLIN component is resolved in two components (A and B) separated by 18.4 pc, with the brightest B located at the northeast. At 200 parsec northeast of B, we detect the resolved out component C, which is the weaker component of the double in the MERLIN image.

All three components in NGC6286 are resolved with sizes between 4 and 10 parsecs, brightness temperatures ranging from 1.02×10^6 K (C) to 8.9×10^6 K (B) and luminosities from 10^{37} to 10^{39} erg/s, hence a few times higher than the RSN in Arp 220.

The fact that they are resolved argues against single RSN. It is plausible that one of the compact components may be due to a clump of supernovae. It is also possible that the brightest components are due to AGN activity.

3.3. Mrk 266

Mrk 266 is a system of interacting galaxies (at a distance $d = 113$ Mpc) showing an interaction region in the middle and tidal tails. Mazzarella et al. (1988) classify the northern as a LINER and the southern galaxy as a Seyfert 2 type. VLA and MERLIN

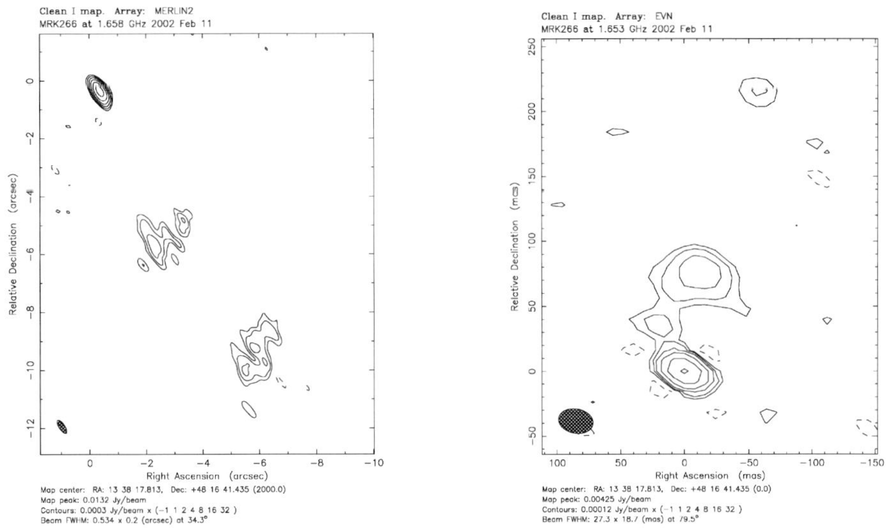

Figure 3. Left: The MERLIN image of Mrk 266 with a restoring beam of 534 × 200 mas. The peak is 13.2 mJy/bm and the contours are drawn at 1, 2, 4, 8, 16, 32 × 0.3 mJy/bm. Right: The EVN image of Mrk 266 restored with a beam of 27 × 18 mas. The peak is 4.2 mJy/bm and the contours are drawn at 1, 2, 4, 8, 16 × 0.12 mJy/bm.

images show at least three co-linear components of which the outer are associated with the two interacting galaxies and the central may be a region of synchrotron emission stimulated by the collision of the two galaxies with the interaction region. With increasing resolution, the two southern components become resolved while the northern (which is the target of our EVN observations) remains compact.

In our MERLIN image (Figure 3), the northern component has a flux density of 21 mJy, consistent with the image of Thean et al. (2001) and a spectral index $\alpha_{1.6\text{–}8\,\mathrm{GHz}} = 0.93$. The higher resolution EVN observations (beam = 21 mas, 107 pc) show a relatively compact (2 × 2 pc) component (3.4 mJy) and arc-like emission which culminates in a more extended (18 × 14 pc) weaker (1.7 mJy) subcomponent ∼41 pc, while a third weak component is visible ∼100 pc further north. Their brightness temperature range from 2.3×10^7 to 3.3×10^5 K and their luminosity from 4×10^{38} to 8.2×10^{38} erg/s.

Combining our EVN image with archival MERLIN 5 GHz data, we find that the brightest, compact component has a steep spectrum of $\alpha_{5\text{–}1.6\,\mathrm{GHz}} = 0.681$, while weaker and more extended component has a rising spectrum of $\alpha_{5\text{–}1.6\,\mathrm{GHz}} = -0.023$.

4. Summary

In the three ULIRGs discussed in this paper, 50–70% of the total flux density is resolved out at parsec scale resolution. The parsec scale structure consists of multiple components with sizes of a few to tens of parsecs, brightness temperatures ranging from 10^5 to 10^7 K and luminosities from 10^{37} to 10^{40} erg/s (UGC 05101). All components are too large to be single radio supernovae and apart from UGC 05101, their luminosities are a few thousand times brighter than Cas A but similar to the supernova 1989J. It is therefore possible that at least some of these components are clustered young radio supernovae (RSN) or supernovae remnants.

It appears that at least some of the high brightness temperature parsec scale emission is due to an AGN, in particular in UGC 05101 whose components are much more luminous than 1989J or the supernovae in Arp 220. This plausibility needs to be explored with further multi-frequency observations as well as X-ray observations.

References

Condon, J.J., Helou, G., Sanders, D.B. and Soifer, B.T.: 1990, *ApJS* **73**, 359.
Condon, J.J., Huang, Z.-P., Yin, Q.F. and Thuan, T.X.: 1991, *ApJ* **378**, 65.
Imanishi, M., Terashima, Y., Anabuki, N. and Nakagawa T.: 2003, *ApJ* **596**, L167.
Lonsdale, C.J., Lonsdale, C.J. and Smith, H.E.: 1992, *ApJ* **391**, 629.
Lonsdale, C.J., Smith, H. and Lonsdale, C.J.: 1993, *ApJ* **405**, L9.
Lonsdale, C.J., Lonsdale, C.J., Smith, H.E. and Diamond, P.D.: 2003, *ApJ* **592**, 804.
Mazzarella, J.M., Gaume, R.A., Aller, H.D. and Hughes, P.A.: 1988, *ApJ* **333**, 168.
Smith, H.E., Lonsdale, C.J., Lonsdale, C.J. and Diamond, P.J.: 1998a, *ApJ* **493**, L17.
Smith, H.E., Lonsdale, C.J. and Lonsdale, C.J.: 1998b, *ApJ* **492**, 137.
Thean, A.H.C, Gillibrand, T.I., Pedlar, A. and Kukula, M.J.: 2001, *MNRAS* **327**, 369.

MOLECULAR GAS IN THE ABELL 262 CLUSTER GALAXY UGC 1347*

THOMAS BERTRAM[1], ANDREAS ECKART[1], MELANIE KRIPS[1]
and JOHANNES STAGUHN[2]

[1]*I. Physikalisches Institut, University of Cologne, Germany; E-mail: bertram@ph1.uni-koeln.de*
[2]*NASA/Goddard Space Flight Center, Greenbelt, Maryland 20771, USA*

(Received 16 April 2004; accepted 15 June 2004)

Abstract. We present the results of ^{12}CO(1-0) and ^{12}CO(2-1) observations on UGC 1347 obtained with BIMA and the IRAM 30 m telescope. UGC 1347 is a member of the Abell 262 cluster. In Abell 262, a nearby spiral rich cluster, the signs of galaxy interaction and therefore the mechanisms which play an important role in galaxy evolution within clusters can be studied with high spatial resolution. Aside from its bright central region, UGC 1347 features a second prominent source at the southern tip of the bar, which has been identified as region with recent enhanced star formation. The CO observations prove the existence of reservoirs of cold molecular gas at the positions of both bright regions.

Keywords: galaxies: ISM, galaxies: evolution, galaxies: interactions

1. Introduction

Within clusters – unlike in samples of isolated objects – the fraction of spirals decreases measurably with redshift (Butcher and Oemler, 1978). This evolution is commonly explained with the interaction of cluster member galaxies with either neighboring galaxies, with the cluster's gravitational potential or with the intracluster medium (ICM). Several clear signs of interaction have been observed. Among them is the HI deficiency of galaxies located close to the cluster center reported by Giovanelli and Haynes (1985), Solanes et al. (2001) and others. It is assumed that the affected objects have lost significant fractions of their atomic gas when passing through the center of their host cluster. Interactions consequently influence not only the overall content of HI but also its distribution within the disk (in the case of Abell 262 this can be seen e.g. in the HI maps of Bravo-Alfaro et al. (1997)). Although interactions lead to an overall decrease of the gas mass and hence to a reduced star formation rate (SFR) in the long term, interactions can cause a temporary increase in the gas density and along with this an increased SFR (Vollmer et al., 2001).

Molecular gas is less influenced by the stripping mechanisms, nevertheless it is worthwhile carrying out CO observations on cluster members. Star formation

*This work was partially supported by the SFB 494.

occurs in cold and dense environments traceable by CO. To learn more about star formation in cluster member galaxies triggered by interaction as well as a possible influence on the distribution of molecular gas, observations with highest spatial resolution (subkiloparsec) are mandatory.

The spiral rich cluster Abell 262 is an ideal target to perform these kinds of investigations. It is located nearby, at $z = 0.0163$ (Struble and Rood, 1999), therefore it allows highest spatial resolution observations with current high angular resolution local mm arrays. Giovanelli and Haynes found galaxies in the vicinity of the center of Abell 262 to be HI deficient. The wealth of spirals allows to study the mechanisms responsible for the evolution of cluster galaxies, which may be still at work in this cluster.

UGC 1347 (Figure 1) is an almost face-on SBc galaxy, about 28′ north of the center of Abell 262, well within the region in which the largest amount of HI deficiency is observed. The difference between the systemic velocity of UGC 1347 (5543 km s^{-1}, Haynes et al. (1997)) and the cluster velocity (4887 km s^{-1}, Struble and Rood (1999)) of ~650 km s^{-1} is explainable with the motion of the galaxy within the cluster. 650 km s^{-1} corresponds to 1.2 σ_{LOS}, the velocity dispersion

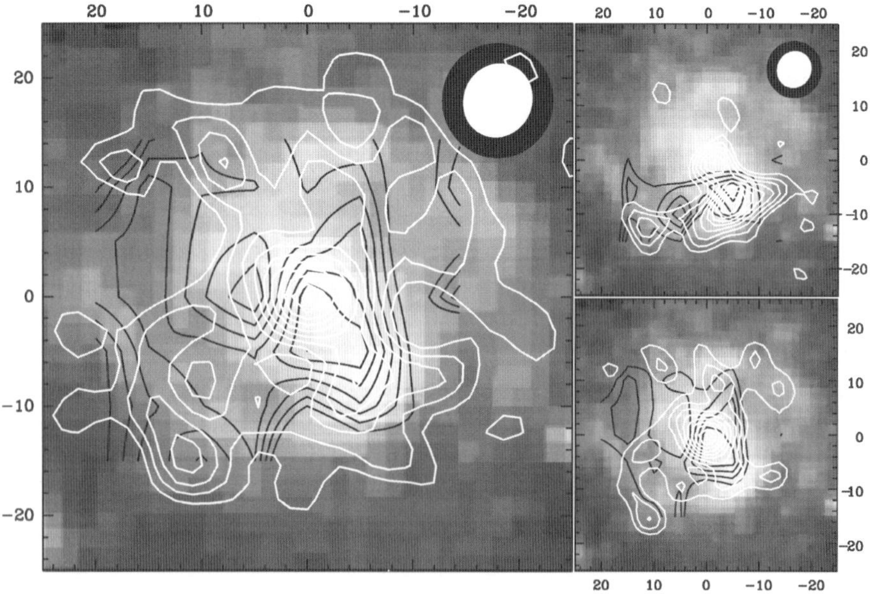

Figure 1. Overlay of the integrated ^{12}CO maps of UGC 1347 obtained with BIMA (^{12}CO(1-0), white contors, peak flux = 17.1 Jy beam^{-1} km s^{-1}) and IRAM 30 m (^{12}CO(2-1), black contours, 8.4 K km s^{-1}) and the Palomar digitized sky survey visible image. The contours represent 10% steps of the peak flux. The CO emission extends over most of the inner disk. Two prominent sources are visible in the visible and the CO maps: the center of the galaxy and a bright star forming region at the southern tip of the bar. Only BIMA channels contributing at least 2σ to the southern and central component were used for the compilation of the integrated flux displayed in the upper right and lower right panel respectively.

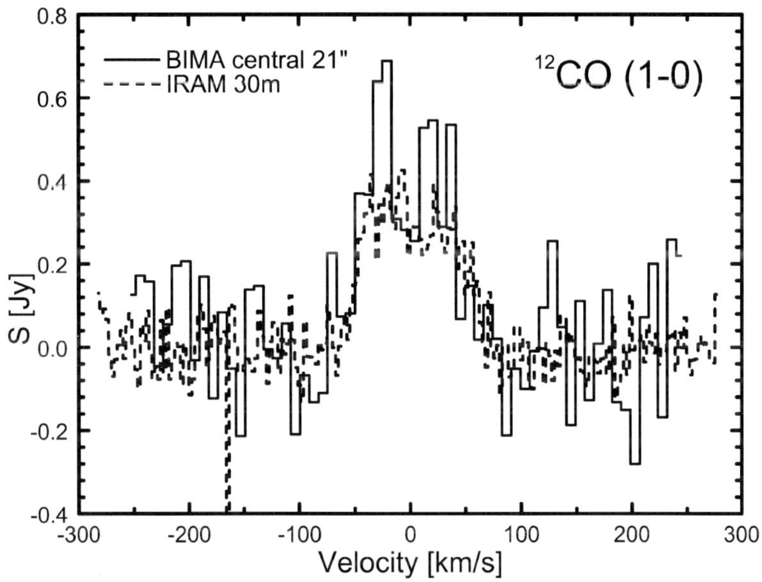

Figure 2. Comparison of two ^{12}CO(1-0) spectra, obtained with the IRAM 30 m telescope and with BIMA. The BIMA flux density was integrated over the inner $21''^2$, the region covered by the IRAM 30 m telescope beam. The spectra indicate that the amount of flux density resolved by the interferometer is negligible.

of the cluster. Initial NIR adaptive optics investigations (Hackenberg et al., 2000) combined with data from other wavelengths have revealed enhanced star formation activity in the nucleus of UGC 1347 and an unusually bright region at one tip of a stellar bar. The NIR colors of the active regions and the disk in total are shown in Figure 4. This activity, UGC 1347's velocity, its HI deficiency although being (in projection) positioned close to the center, and its HI morphology indicate that UGC 1347 is passing for one of the first times through the cluster center and being subject to the effects of galaxy interaction.

High resolution CO observations have been carried out with BIMA (^{12}CO(1-0)) and with the IRAM 30 m telescope (^{12}CO(2-1)). The results of these observations are presented in the following. The compact C configuration of the BIMA array and an on-source integration time of 257 minutes resulted in a good uv coverage and prevented flux resolution (see Figure 2).

2. Molecular Gas in UGC 1347

Figure 1 shows that the CO emission extends over most of the inner disk of UGC 1347. It was possible to identify CO bright regions that coincide with the nucleus and with the NIR bright southern component. Other than in the NIR, where

TABLE I

Parameters measured or derived from CO data[a]

	$I_{CO(1-0)}$ [K km s^{-1}]	$I_{CO(2-1)}$	Θ_S [″]	R	m_{H_2} $M_\odot$
Nucleus	40.1 ± 1.3	8.4 ± 1.0	<6.6	0.37	$4.7 \cdot 10^7\ h^{-2}$
Southern comp.	$24.1^{+5.5}_{-1.6}$	40.0 ± 0.6	5.4	0.33	$9.4 \cdot 10^7\ h^{-2}$
Total galaxy	4.56 ± 0.29		18.9		$5.7 \cdot 10^8\ h^{-2}$

[a]The source sizes are intrinsic sizes obtained under the assumption of Gaussian brightness distributions and Gaussian beams (FWHM: 6.6″ for CO(1-0) and 10.5″ for CO(2-1)). The $I_{CO(2-1)}/I_{CO(1-0)}$ line ratio R was derived for the bright components by applying the correction factor $f_{gauss} = \frac{1+(10.5''/\Theta_S)^2}{1+(6.6''/\Theta_S)^2}$ to the observed line ratios. The molecular mass was estimated using $N(H_2)/I_{CO} = 4 \cdot 10^{19}$ cm^{-2} (K km s^{-1})$^{-1}$ for the nuclear component and $2 \cdot 10^{20}$ cm^{-2} (K km s^{-1})$^{-1}$ for the southern component and the total galaxy. $H_0 = 100\ h$ km s^{-1} Mpc^{-1} has been used.

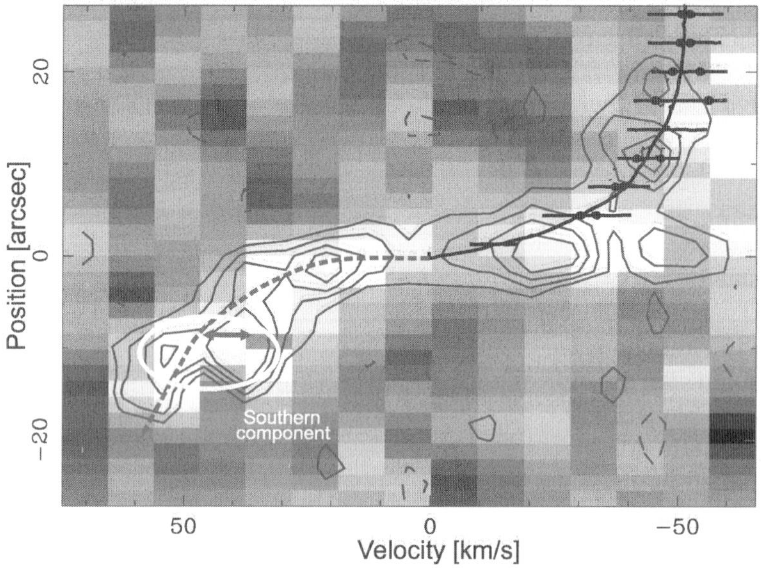

Figure 3. ^{12}CO(1-0) Position-Velocity diagram of UGC 1347 for a position angle of 20°. Superimposed in the top right corner is the Hα rotation curve of Amram et al. (1994). In the lower left quarter, the southern component is clearly visible and appears blue shifted with respect to the disk radiation at that position by approximately 10 km s^{-1}.

it was found to be compact, the southern source is extended in the CO emission. The CO line parameters that have been measured and derived from the CO observations are listed in Table I. In both regions the CO lines are optically thick – under normal conditions an indication for cold, subthermally excited and dense molecular gas. The determination of the overall H_2 content using CO emission (especially optically thick line emission) as tracer has to be handled with care. The common approach

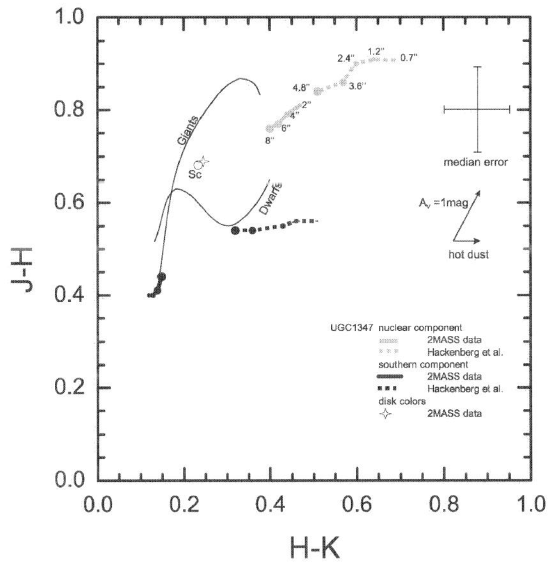

Figure 4. NIR two-color diagram of UGC 1347 based on 2MASS and Hackenberg et al. data. The dot sizes represent diffenent aperture diameters. The nuclear component is in agreement with a normal stellar disk population that is reddened and affected by hot dust. The disk of UGC 1347 in the diagram is at the position expected for normal Sc galaxies. In the case of the unresolved southern component the 2MASS data with its poorer resolution is probably diluted by the light of young stars in the star forming region.

of using a 'standard' $N(H_2)/I_{CO}$ conversion factor X derived from galactic observations has the shortcoming of disregarding the dependency of X on the metalicity of the region of interest. Since information on the metalicity in UGC 1347 is not available, the estimate on the molecular gas mass m_{H_2} has to remain approximate. To account for the higher metalicity in galaxy centers, a lower conversion factor than for the other two regions has been used for the central region, as suggested in Israel (2001).

The Hα velocity field and rotation curve of UGC 1347 was observed by Amram et al. (1994). They derived an approximate inclination of 30° and a position angle of 20°. As shown in Figure 3, the Hα rotation curve is in good agreement with the BIMA CO rotational parameters. Clearly visible in the position-velocity diagram is the southern component. It appears to be blue shifted by approximately 10 km s^{-1} with respect to the underlying disk. This may be explained with a model of an an expanding gas shell.

An upper limit for the dynamical mass in the unresolved center can be derived by considering the maximum velocity measured at the center position in the position-velocity diagram of ~50 km s^{-1}. Accounting for an inclination of 30° (resulting in $v_{\max} = 100\,\mathrm{km\,s^{-1}}$) and assuming a maximum radius of below 1.5″ and corotation in the central region, the total mass in this region has to be below $8.3 \times 10^8 h^{-1}\, M_\odot$.

processing is required beyond observation and data reduction. The simplicity of the OH Zeeman diagnostic arises from the fact that, unique amongst the molecular species that generate strong masers, OH has a Zeeman effect of hyperfine structure (Dousmanis et al., 1955). The important point is that, for typical magnetic fields found in star-forming regions (a few mG), most OH maser transitions have Zeeman components split by considerably more than the typical Doppler width of the lines themselves and, under these conditions, the magnetic field is proportional to the splitting. Therefore, with a few provisos, an observation of a Zeeman pair yields a measured component splitting, and a corresponding magnetic field strength. As an example of what can be achieved in a rich source, a magnetic field contour map of W3(OH) has been constructed, based on 211 OH maser Zeeman pairs from 1.7-, 6.0- and 13-GHz data (Baudry and Diamond, 1998; Desmurs et al., 1998; Wright et al., 2004).

The diagnosis of magnetic field strengths from OH masers is not *quite* straightforward, and a few cautionary points need to be noted. Firstly, it is unwise to compute magnetic fields from apparent Zeeman patterns in single-dish spectra: it is necessary to show that the components of a Zeeman pattern are spatially coincident on a scale comparable with their resolved size, which typically involves observations with VLBI resolution (Lo et al., 1975). Secondly, the OH maser transitions fall into two broad classes: main lines, in which the upper and lower hyperfine levels have the same F-quantum number, and satellite lines where they do not. Main lines have an unambiguous splitting for a given magnetic field, but satellite lines have a more complex pattern because the upper and lower energy levels are split by different amounts for a given magnetic field. For example, the satellite lines of the rotational ground state (the 1612- and 1720-MHz transitions) have six σ transitions, which can be sorted into three $\sigma+/\sigma-$ pairs, each with its own velocity splitting for a given magnetic field. Therefore, to derive a field from the splitting of a left-right Zeeman pair in a satellite line, it is necessary to know which of the three possibilities is being observed; in the case of a single pair, it is generally impossible to know this for certain, but it can often be inferred by comparing fields derived from the satellite line with those derived from the closest main-line Zeeman pairs, and assuming that the fields should be similar. Thirdly, the Zeeman splittings per mG of field generally decrease moving from the OH ground state to the more exotic excited-state OH masers, such as 13.44 GHz. As the splittings become smaller, the left and right elliptically polarized components tend to be more overlapped, and the approximations used in the well-split case, including taking the derived magnetic field to be proportional to the total magnetic field strength, are less likely to be valid.

The sense of the magnetic field, that is whether the head of the field vector points generally towards or away from the observer, can be determined by whether the higher frequency component of a Zeeman pair is left- or right-elliptically polarized. Recovery of the full vector magnetic field requires considerable additional modelling, and requires linear polarization information. It is therefore wise, if any attempt at vector reconstruction is intended, to perform observations in full polarization mode. A successful use of such modelling has been the interpretation of spatially superimposed crossed planes of polarization (Hutawarakorn and Cohen, 1999). It can be shown, using a magnetohydrodynamic disc and outflow model (Uchida and Shibata, 1985) and a detailed maser simulation, that apparent overlay of features with polarization planes crossed at angles of ~90 degrees arises naturally if star-forming region maser sources are modelled as inclined rotating discs. The interpretation of polarized OH maser data from W75N appears in Gray et al. (2003) and requires that observed masers can originate from different depths in the disc, relative to its equatorial plane. Similar methods have also been applied to linearly polarized H_2O masers in W3 IRS5 (Imai et al., 2003). Megamasers appear insufficiently polarized to yield magnetic field information through the OH Zeeman effect (Killeen et al., 1996), so at present the technique is restricted to Galactic sources.

3. Proper Motions and Cloud Dynamics

Proper motion measurements rely on the persistence of a large proportion of maser spots over timescales which allow time-series measurements to detect significant motion in the plane of the sky. Assuming that the masers follow the motions of their host gas, it is possible to use the maser spots as markers to follow the dynamics of the source. The annual motion of a maser spot is $210 v_{\mathrm{k}}/d_{\mathrm{pc}}$ μas for an object at a distance of d_{pc} parsecs, with tangential velocity v_{k} km s^{-1}. Observations that can detect proper motion therefore need to be of high-resolution interferometric type, usually of VLBI scale. Information about the radial component of the velocity is available from Doppler measurements, so it is possible to reconstruct fully the 3-D velocity field of the source.

The velocity-field information provided by proper motion observations is usually used in a statistical manner to endorse, or to rule out, various dynamical models of the source. In Table I, I show a selection of proper motion observations, making use of masers from various molecular species, and indicating the broad conclusions drawn about the internal motions of the sources. Also shown are the, often long, intervals between the sets of observations.

An example of a proper motion analysis involves OH and H_2O masers in W3(OH)TW (the Turner-Welch object). Here the motion of the water masers indicates general outflow. Strangely, in this source, the OH masers share the outflowing motion of their H_2O counterparts, which is not typical behaviour for OH masers. The Turner-Welch object is almost certainly younger than the neighbouring W3(OH): its HII region is still obscured, so we could be observing a new class of OH masers associated with cometary shocks, and found in the earliest evolutionary stages of massive star formation.

TABLE I

A selection of proper motion observations

Object	Species	ν (GHz)	Interval (yr)	Ref.	Notes
W3(OH)	CH_3OH	12.2	6	(a)	Keplerian disc?
W3(OH)	OH	1.7	Note	(b)	cannot be just rotation
Ceph A	H_2O	22.2	7	(c)	exp. into rotating disc
W51-IRS2	H_2O	22.2	19	(d)	bow shock
W75N	H_2O	22.2	1	(e)	jet + shell
W3(OH)TW	OH	1.7	8	(f)	OH linked to H_2O
NGC2071IR	H_2O	22.2	0.33	(g)	outflows $\perp$ to disc

(a) (Menten et al., 1988; Moscadelli et al., 1999)
(b) (Reid et al., 1980) and see *Note*
(c) (Torrelles et al., 1998; Gallimore et al., 1998)
(d) (Schneps et al., 1981; Eisner et al., 2002)
(e) (Torrelles et al., 2003)
(f) (Argon et al., 2003)
(g) (Seth et al., 2002)
Note: There are several observing epochs for OH masers in W3(OH), spanning at least the years 1976–1996.

The main limitations on proper motion techniques are attached to the timescales required: for OH masers intervals approaching a decade are required to get accurate motions, though a year or so is often acceptable for the faster-moving CH_3OH and H_2O masers. Assuming that an observer is prepared to wait a very long time for results, the quality of the information extracted then depends on the variability of the maser spots: for OH in Galactic star-forming regions, a large proportion are still identifiable in observations separated by more than a decade. Water masers vary more rapidly, so over a similar timescale, many more spots would be expected to lose their identity.

Proper motion analysis has also been used successfully in extragalactic megamasers: in particular, for the specialist application of distance determination. To find distances requires the velocity gradient diagnostic in addition to proper motions (see Section 4). The best example is NGC4258 (Herrnstein et al., 1999); the maser distance has recently been compared with a distance derived from Cepheid variables (Caputo et al., 2002) and the agreement is very good. Another interesting recent development is the discovery of microstructures (Torrelles et al., 2001) in which well-correlated velocity patterns can be tracked at scales down to 1 AU.

4. Velocity Gradients and Binding Masses

Class II methanol maser clusters in Galactic star-forming regions often appear as lines of maser spots in the plane of the sky, with a clear Doppler velocity gradient along the line (Norris et al., 1998). Similar arrangements may be found, though less frequently, in OH and H_2O masers. In the particular case of OH masers, velocity gradients can often be obscured by the Zeeman effect and, in any test for a velocity gradient, it is important to use demagnetised velocites derived from the midpoints of the Zeeman pairs. A well-defined North-South velocity gradient was recently found for OH masers in W3(OH) based upon demagnetised velocities (Wright et al., 2004).

Initially, the discovery of a velocity gradient allows only the reconstruction of a map of the radial velocity field in the source but, with certain assumptions about the geometry of the source, additional information can be gleaned: The usual model applied is that the masers are distributed in a rotating disc, seen close to edge-on. In this case, the Keplerian formula,

$$dv/dx = (GM/r^3)^{1/2} \cos i \tag{1}$$

can be applied. Using the velocity gradient dv/dx, and an estimate of the inclination angle, i, the mass-density of the source can be calculated and, if the size of the source is known, the binding mass can be found. The velocity-gradient diagnostic has now been used on all scales from protoplanetary discs to kpc-size sources in megamasers. I show a sample of observations and derived masses in Table II.

TABLE II

A sample of mass measurements derived from velocity gradients in maser groups

Object	Object type	Species	ν (GHz)	Mass ($M_\odot$)
NGC4258	megamaser	H_2O	22.2	2.2×10^7 (Haschick et al., 1994)
NGC4258	megamaser	H_2O	22.2	3.9×10^7 (Herrnstein et al., 1999)
IIIZw35	megamaser	OH	1.667	5.2×10^6 (Diamond et al., 1999)
IIIZw35	megamaser	OH	1.667	1.0×10^7 (pihlstroem et al., 2001)
MKN273	megamaser	OH	1.7	1.5×10^8 (Yates et al., 2000)
NGC7538	massive SFR	CH_3OH	12.2/6.7	31 (Minier et al., 1998)
NGC6334F	massive SFR	OH	1.7	25 (Zheng, 1989)
W3(OH)	massive SFR	OH	1.665	27 (Wright et al., 2004)
W75N	massive SFR	CH_3OH	12.2/6.7	4.4 (Minier et al., 2000)
MonR2	massive SFR	CH_3OH	12.2/6.7	74.6 (Minier et al., 2000)
G29.95-0.02	massive SFR	CH_3OH	12.2/6.7	30.8 (Minier et al., 2000)
NGC7538-IRS1	compact HII	H_2CO	4.83	~1 (Hoffman et al., 2003)
NGC2071	protoplanetary	H_2O	22.2	1 (Torrelles et al., 1998)
Orion	protoplanetary	H_2O	22.2	15 (Abraham and Vilas Boas, 1994)

Notes: Data from Minier et al. (2000) assume Case 2.
Mass for NGC7538-IRS1 computed from gradient in Hoffman et al. (2003).

5. Co-Propagation and Physical Conditions

The computation of the physical parameters of a gas (such as its kinetic temperature, density and composition) from measurements of the relative strengths of spectral lines has a long and important history in astrophysics, both for stellar atmospheres and the interstellar medium. Maser lines can, in principle, be used in a similar manner, but the problem is much more complicated: a first requirement is that two or more maser lines must be identified as co-propagating through the same column of gas. To know whether this is so, the absolute positions of the candidate maser spots must overlap to within a scale comparable to the mean of their resolved sizes. Such knowledge requires externally phase-referenced observations with MERLIN or VLBI networks. Even if candidate masers (possibly from different molecular species) do overlap in velocity and position, the interpretation of their relative intensity requires detailed radiation transfer theory unless the masers are unsaturated, when a knowledge of the background intensities is necessary.

Maser co-propagation has been detected between OH 1720- and 4765 MHz masers in W3(OH) with MERLIN (Gray et al., 2001) and with the higher resolution of the VLBA (Palmer et al., 2003). There are additional instances of co-propagation between CH_3OH 4.6- and 12.2-GHz masers, also in W3(OH). The use of co-propagating masers to recover physical conditions is still in its infancy. Despite multi-species propagation theories (Cragg et al., 2002), the main requirement is still for intelligent optimization software which is capable of rapidly performing the inverse radiation transfer problem for masers.

6. Long-Term Variability and Turbulence

When considering variability, I have arbitrarily decided to divide the subject into two classes: I take long-term variability to occur on timescales longer than the typical light-propagation time through the maser, and short-term variability to operate on shorter timescales than this. I have also considered only intrinsic variability: I have ignored such effects as scintillation which result from the properties of the ISM between the maser source and the Earth.

Masers in star-forming regions are often modelled as a chain of velocity-coherent regions of gas, possibly with magnetic velocity compensation (Cook, 1966). Under this scheme, variability results from bulk gas flows and turbulence in the source, which move parts of the maser column in and out of velocity coherence along a particular line of sight. For most sources in star-forming regions, the form of the variability is some complicated function of the parameters of turbulence in the source, but as there is little prospect of being able to do the inverse problem analytically, the turbulence parameters (intensity and spectral index for example) need to be found by parameter-space searches, aiming for a best fit to observations from a variety of theoretical predictions. In the future, it may be possible to relate turbulence properties directly to the masses and ages of star-forming regions.

Single-dish observations are valuable in studying variability. A good example is the 13 yr study of 14 star-forming regions (Brand et al., 2003) which has shown that objects with higher far-infrared luminosity have stronger and more stable masers: there appears to be a lower limiting luminosity of $\sim 130\,L_{\odot}$ below which sources are incapable of supporting H_2O masers. Excellent studies of CH_3OH maser variability in star-forming regions, at 6.7 and 12.2 GHz, have also been performed (Goedhart and Gaylard, 2002). Unexpectedly, Goedhart and Gaylard discovered three types of variability: aperiodic (including flaring), quasi-periodic, and periodic variability. The latter form is difficult to explain in terms of turbulence, but could be linked to binarity of the source or a precessing jet interacting with the maser zone.

High-resolution observations give us variability information about individual maser gain paths, and provide the best chance of converting the observed variability into physical parameters of the turbulence (Kolmogorov intensity index and Gaussian amplitude width). For example, observations of CH_3OH masers (Johnston et al., 1997) have been modelled in terms of turbulence, producing very realistic-looking synthetic VLBI maser spots for the best-fit turbulence parameters (Sobolev et al., 1998; Sobolev et al., 2003). I note that it is now possible to relate transport coefficients, such as the turbulent viscosity and diffusivity, to observational parameters such as the variability timescales and spot sizes of masers, via the Strouhal number (Lekht et al., 1999).

7. Short-Term Variability: Instabilities and Coherence

Variability on such short timescales (excluding external scintillation effects) must result from the processes of maser amplification and saturation themselves. The radiative transfer equation for maser amplification in the time-dependent case can be subject to instabilities, and such effects have been predicted for H_2O masers (Scappatici and Watson, 1992). Instabilities of this type produce variability in the maser output on timescales $\sim L/c$, where L is the gain-length of the maser. For water masers this translates to hours or even minutes. These effects should not be confused with the *time-independent* instabilities which are not predicted by all theories (Elitzur, 1993; Watson, 1994; Gray, 2003).

Astrophysical masers should also possess some residual coherence over and above what is expected from an incoherent source. Theory (Gray and Bewley, 2003) shows that phase-coherent effects can exist over a frequency range which is only of the same order as the homogeneous width of the maser line: essentially its natural width, invariably dwarfed by the inhomogeneous linewidth that reflects the temperature of the masing gas and the intensity of small-scale turbulence. An early experiment that attempted to detect the residual coherence of astrophysical

Uchida, Y. and Shibata, K.: 1985, *PASJ* **37**, 515.
Watson, W.D.: 1994, *ApJ* **424**, L37.
Wright, M.M., Gray, M.D. and Diamond, P.J.: 2004, *MNRAS* **350**, 1253.
Wright, M.M., Gray, M.D. and Diamond, P.J.: 2004, *MNRAS* in press.
Yates, J.A., Richards, A.M.S., Wright, M.M., Collett, J.L., Gray, M.D., Field, D. and Cohen, R.J.: 2000, *MNRAS* **317**, 28.
Zheng, X.-W.: 1989, *ChJSS* **9**, 87.

SPECTRAL LINE AND CONTINUUM RADIATION PROPAGATION IN A CLUMPY MEDIUM

JOHN CONWAY[1], MOSHE ELITZUR[2] and RODRIGO PARRA[1]

[1]*Onsala Space Observatory, Onsala, Sweden; E-mail: jconway@oso.chalmers.se*

[2]*Department of Physics and Astronomy, University of Kentucky, KY, USA*

(Received 16 April 2004; accepted 15 June 2004)

Abstract. We discuss the propagation of spectral line and continuum radiation in a clumpy medium and give general expressions for the observed absorption or emission from a cloud population. We show that the affect of the medium clumpiness can usually be characterised by a single number multiplying the mean column opacity. Our result provides a simpler proof and generalization of the result of Martin et al. (1984). The formalism provides a simple way to understand the effects of clumping on molecular line profiles and ratios; for example, how clumping effects the interpretation of $^{13}CO(1\text{–}0)$ to $^{12}CO(1\text{–}0)$ line ratios. It also can be used as a propagation operator in physical models of clumpy media where the incident radiation effects the spectral line emissivity. We are working to extend the formalism to the propagation of masers in a clumpy medium, but in this case, there are special difficulties because formal expectation values are not characteristic of observations because they are biased by rare events.

Keywords: molecular lines, spectra

1. Introduction

The problem of the propagation of radiation in a non-uniform medium is a general one in astrophysics, which arises in many different contexts. We consider here the common case that the medium consists of clumps or clouds with a low volume filling factor, but potentially high covering factor. These clumps or clouds may possess a range of properties. This paper gives a summary of our derivation and results, a more detailed proof will be published elsewhere (Conway et al., 2005, in preparation).

The specific problem we address is the expected response of a telescope beam observing a low filling factor clumpy region where the beamwidth is much larger than the typical cloud diameter. In this case, we can show that that expectation value of the beam average equals the ensemble expectation value over many realisations of observations along the line of sight (LOS). We first (see Section 2) analyse the case of continuum (e.g. dust) clouds absorbing background radiation, then spectral line absorption (Section 3). In Section 4, we consider spectral line emission and apply the formalism to the interpretation of CO line ratios. Finally, in Section 5, we draw conclusions and describe future work.

Astrophysics and Space Science **295:** 319–324, 2004.

2. Continuum Absorption Radiation Transfer

Let us consider continuum radiation transfer in a clumpy medium of dust clouds. First assume that all clouds/clumps are identical. Assume further that radiation I entering a clump emerges as $I \cdot R$, where the clump response is R. In terms of cloud opacity $R = \mathrm{e}^{-\tau}$. Hitting successively k clumps that all have the same response factor produces the aggregate response R^k. Implicit in this assumption is that the interaction with the radiation does not change the clump response factor R. It can be shown that provided the clump volume filling factor is small, the probability of getting k clumps along a given line of sight always obeys a Poisson distribution, characterised by one number, the mean $\bar{N}$ along a LOS. In this case of one clump type, it can be shown that the mean response is

$$\bar{R} = \exp^{-t}, \qquad t = \bar{N}(1 - \mathrm{e}^{-\tau}) \tag{1}$$

as previously derived by Natta and Panagia (1992).

Next consider the case of many types of clumps with response factors R_i (these could be clump types having different optical depths) each obeying Poisson statistics. If the fraction of clumps of each type is f_i then (see Conway et al., 2005, in preparation)

$$\bar{R} = \exp(\bar{N}[R_{\mathrm{av}} - 1]), \quad \text{where } R_{\mathrm{av}} = \sum_i f_i R_i \tag{2}$$

This means that the problem of a mixture of cloud optical depths (τ_i) is equivalent to that of the single cloud type having an average τ_{av} obtained from

$$R_{\mathrm{av}} = \mathrm{e}^{-\tau_{\mathrm{av}}} = \sum_i f_i\, \mathrm{e}^{-\tau_i} \tag{3}$$

We can generalize the result in Eq. (2) to the case where the clump type varies continuously. For instance, if the clump type is described by continuous variables t and r (which may correspond for instance to peak optical depth and distance from the centre of clouds which have spatial structure) then we have

$$R_{\mathrm{av}} = \int\int R(r,t)\eta(r,t)dr\,dt = \int\int \mathrm{e}^{-\tau(r,t)}\eta(r,t)dr\,dt \tag{4}$$

and likewise for any number of dimensions of continuous parameters, where η is the distribution function of clump types.

3. Line Radiation Transfer – Absorption

In the spectral line absorption problem, we aim to calculate the expectation value of a frequency-dependent absorption response function given a population of clouds. This cloud population has in general a distribution of central LOS velocitie, a range of spectral profiles, and spatial structures. Observed absorption lines are generally published as a function of velocity v, the response at velocity vR(v) being understood to represent the response at observing frequency $\nu = \nu_0/(1+\mathrm{v}/c)$ where ν_0 is the line's rest frequency. Let us concentrate on calculating the response at a particular velocity v, $R(v)$. In principle, clouds having all possible centre velocities v' contribute to the spectral response at velocity v. The centre velocity of each cloud v' can simply be thought of as yet another continuous variable describing the cloud properties, so the full distribution function becomes $\eta(r, t, v')$. For spectral line clouds, the response function is $R_v(r, t, v')$ which if cloud types are defined to have fixed spectral profiles which do not vary with centre velocity can be written $R(r, t, v - v')$. Hence, applying Eqs. (2) and (4), we obtain

$$\bar{R}(v) = \exp(\bar{N}[R_{\mathrm{av}}(v) - 1]) \tag{5}$$

where

$$R_{\mathrm{av}}(v) = \int\int\int R(r, t, v - v')\eta(r, t, v')dv'drdt \tag{6}$$

As before $\bar{N}$ is the mean number of clouds (at any velocity) per LOS. $R_{\mathrm{av}}(v)$ can be interpreted at each r, t value as a convolution of the cloud velocity profile for this r, t, $R(r, t, v)$ with the velocity distribution function of the cloud population $\eta(r, t, v)$ followed by an integration over all r, t (and any other continuous cloud parameters).

If the fractional distribution of clouds of different kinds is independent of velocity, then $\eta(r, t, v) = f(r, t)q(v)$ is a separable function. In this case we can define a mean cloud spectral response

$$R_{\mathrm{s}}(v) = \int\int R(r, t, v)f(r, t)drdt \tag{7}$$

In this separable case, the expectation value of the emerging spectrum is

$$\bar{R}(v) = \exp(-\lambda(v) * C(v)) \tag{8}$$

In this expression, $*$ denotes convolution, $\lambda(v) = \bar{N}q(v)$ describes the cloud velocity distribution; more precisely defined as the mean number of clouds per unit velocity per LOS. The quantity $C(v) = 1 - R_{\mathrm{s}}(v) = \langle 1 - \mathrm{e}^{-\tau(v)}\rangle$, where $\tau(v)$ is the opacity profile of each clump type, and the brackets denote an average over the clump population. The $C(v)$ can be interpreted as the mean cloud emission profile. In the normal case that the $\lambda(v)$ distribution is much broader in velocity than $C(v)$, we obtain the good approximation that $\bar{R}(v) \approx \exp(-C_A\lambda(v))$, where $C_A = \int C(v)dv$ has dimensions of velocity and can be considered the equivalent width of the mean cloud profile.

Another way to view the effect of clumping is to note that

$$\bar{R}(v) \approx \exp(-K\,\bar{\tau}_{\rm tot}(v)) \tag{9}$$

where $\bar{\tau}_{\rm tot}(v) = \lambda(v) * \langle\tau(v)\rangle$ is the mean line of sight opacity and K is a factor taking into account clumping which is

$$K = \frac{\int \langle 1 - e^{-\tau(v)}\rangle dv}{\int \langle\tau(v)\rangle dv} \tag{10}$$

Remarkably, we find that the effect of an arbitrarily complex clump distribution can usually be reduced to a single number, K. This factor K reduces the effective opacity of a clumpy medium compared to a smooth gas of the same mean column density. Martin et al. (1984) partially derived our results, but their derivation was less direct and the complete generality of the result was not stated.

4. Emission Line Spectra

The expressions for opacity derived earlier can be adapted (Conway et al., 2005, in preparation) to predict the emissivity in the case of a population of clouds with a mixture of excitation temperatures and statistical properties along the line of sight. Here we discuss only the simplest case that all clouds have the same excitation temperature $T_{\rm e}$. In this case, the observed spectral brightness temperature is

$$T(v) = T_{\rm e}[1 - \bar{R}(v)] = T_{\rm e}[1 - \exp(-K\,\bar{\tau}_{\rm tot}(v))] \tag{11}$$

To illustrate the impact of clumping on observed line profiles and line ratios we show an example in Figure 1 involving the ^{12}CO(1–0) and ^{13}CO(1–0) lines. In this example, the cloud population consists of only one clump type with a Gaussian opacity velocity profile $\tau(v)$, and a much broader Gaussian $\lambda(v)$ cloud velocity distribution. The expectation value of the total opacity along each LOS $\bar{\tau}_{\rm tot}(v) = \lambda(v)*\tau(v)$ is therefore also a Gaussian whose peak we term the 'peak total opacity'.

The different panels for Figure 1 have different combinations of the ^{12}CO(1–0) peak total and peak cloud opacities. The ratio of intrinsic ^{12}CO(1–0) and ^{13}CO(1–0) opacities is set to 60 so that in all the panels of Figure 1 the ^{13}CO(1–0) total and cloud opacities are $\ll$1 and hence are optically thin. In all cases, therefore, the ^{13}CO(1–0) line profiles are Gaussian with peak value (in units of $T_{\rm e}$) equal to

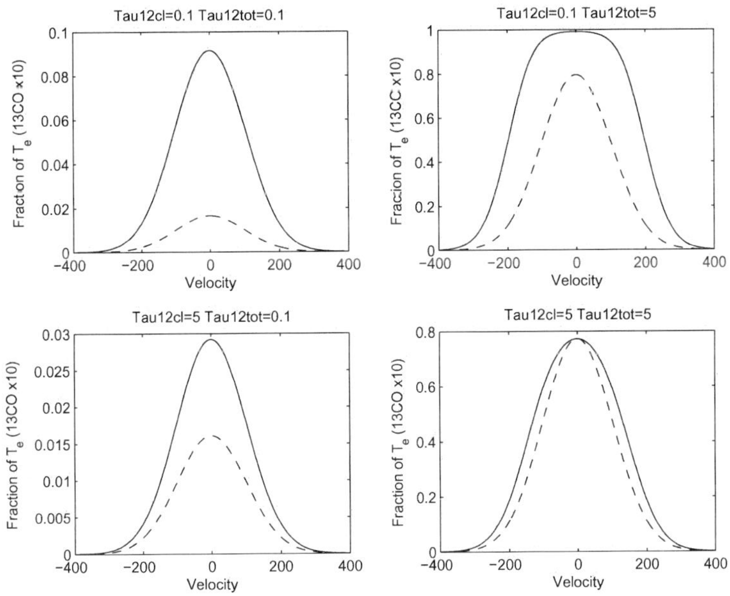

Figure 1. Example ^{12}CO(1–0) and ^{13}CO(1–0) line profiles (shown respectively by solid and dashed lines) for different combinations of cloud and mean column opacities. To make them more visible the ^{13}CO(1–0) profiles have been multiplied by 10. The intrinsic ratio of opacity in the two transitions is assumed to be 60. Both the internal cloud opacity profiles of the clouds in the two transitions and the cloud velocity distribution are assumed to be Gaussian with the latter having a FWHM 10 times larger. The top row shows the case of low opacity in each ^{12}CO(1–0) cloud, and a varying mean *total* ^{12}CO(1–0) opacity. The bottom row shows the case of high opacity ^{12}CO(1–0) clouds. Note the different vertical scales of each plot. Only when the total and cloud ^{12}CO(1–0) opacities are both less than one is the observed ^{12}CO(1–0) to ^{13}CO(1–0) ratio close to the intrinsic value of 60.

the total ^{12}CO(1–0) opacity divided by 60. In contrast, the ^{12}CO(1–0) line profiles and their ratio to the ^{13}CO(1–0) lines depend critically on the clump opacity which affects K. Only for the case in the top left panel, which has both a low total and low clump opacity, is the observed line ratio equal to the intrinsic value. If either or both total and clump opacities are larger than 1, then a smaller line ratio is observed, but it is impossible just from the line ratio at line centre to distinguish the different cases.

An important consequence of Eq. (9) is that when the cloud velocity distribution is much broader than individual clouds it is in principle impossible from the ^{12}CO(1–0) and ^{13}CO(1–0) spectra alone to separate the effects of clumping and variations in the intrinsic ^{12}CO(1–0)/^{13}CO(1–0) opacity ratio. In going from panels on the top row with low opacity clumps to the bottom row with large clumps the ^{12}CO(1–0) clump factor K is reduced. The resulting spectra are the same as if there was no clumping but the intrinsic ^{12}CO(1–0)/^{13}CO(1–0) opacity ratio was reduced.

5. Conclusions and Future Work

We have derived general expressions for both continuum and spectral line radiation propagation in a clumpy medium. These expressions hold provided only that the clumps have a small volume filling factor and that the local emissivity is not itself affected by the local line spectra. Equations (8)–(10) provide simple expressions that can be used by observers to interpret observed line profiles. In Section 4, we illustrated the application of this equation to CO line emission profiles from a clumpy medium. Although the results presented in Section 4 are not new, our equations provide a way to understand line profiles from clumpy media in a systematic and quantitative way.

We are working to extend our formalism to predict variances and higher-order statistics of spectra from clumpy media. We also hope to use our formalism as a propagator term in physical spectral line models of clumpy media. In general, in such models, emissivity will depend on the local spectrum so such models must be iterative. Finally, we are actively working (see Parra et al. these proceedings) on the problem of maser propagation in a clumpy medium. Formally, exactly the same equations apply as in the thermal case, the only difference is the use of negative opacities. However, as well as physical complications of maser saturation there is a more subtle effect that even for unsaturated masers the formal means predicted by the equations are not good predictors of observations. The reason for this is that these mean values are dominated by events causing very large amplification which are also very improbable, such that they never occur within a typical source.

References

Martin, H.M., Hills, R.E. and Sanders, D.B.: 1984, *MNRAS* **208**, 35.
Natta, A. and Panagia, N.: 1984, *ApJ* **287**, 228.

A THIN RING MODEL FOR THE OH MEGAMASER IN IIIZw35

RODRIGO PARRA[1], JOHN CONWAY[1] and MOSHE ELITZUR[2]
[1]*Onsala Space Observatory, Sweden*
[2]*Department of Physics and Astronomy, University of Kentucky, USA;*
E-mail: rodrigo@oso.chalmars.se

(Received 16 April 2004; accepted 15 June 2004)

Abstract. We present a model for the OH megamaser emission in the starburst galaxy IIIZw35. The observed diffuse and compact OH maser components in this source are explained by a single phase of unsaturated clumpy gas distributed in a thin ring structure and amplifying background continuum. We emphasize the importance of clumpiness in the OH masing medium, an effect that has not been fully appreciated previously.

The model explains why multiple bright spots are seen only at the ring tangents while smoother emission is found elsewhere. Both the observed velocity gradients and the line to continuum ratios around the ring enquire a geometry where most of the seed photons come from a continuum emission which lies outside the OH ring. To explain both the OH and continuum brightness, free-free absorbing gas is required along the ring axis to partially absorb the far side of the ring. It is proposed that the required geometry arises from an inwardly propagating ring of starburst activity.

Keywords: starburst, maser

1. Introduction

Extragalactic OH megamasers usually present both extended diffuse emission and very compact and strong features. One of the clearest cases of an OH megamaser showing both compact and diffuse masers is IIIZw35. Two groups of compact masers were detected in VLBI (Trotter et al., 1997; Diamond et al., 1999) recovering nearly half of the total emission seen at MERLIN scales (Montgomery and Cohen, 1992). Aiming to determine the location of the missing diffuse component MERLIN+EVN observations were conducted on this source (Pihlström et al., 2003). These observations revealed the best example of a rotating OH maser ring that has yet been found.

It has been claimed that the two-phase characteristic of the maser features observed toward IIIZw35 is due to two coexisting phases of the OH gas (Diamond et al., 1999), one producing the smooth emission and the other responsible for the very strong and compact features. According to this hypothesis, and making the usual assumption that the diffuse component maser is operating in the unsaturated regime, the Line to Continuum Ratio (LCR) of each phase is exponentially related

Astrophysics and Space Science **295:** 325–330, 2004.

to the path length through the amplifying gas. In the particular case of the smooth component in IIIZw35, the LCR in the east/west and north/south regions has been observed to be around 10 and 1000, respectively. Additionally, the ratio of the path lengths between these regions is estimated to be approximately 8 therefore, the ratio of their LCRs should be in the order of 10^7. Instead in order to match the observations, a clumpy medium is proposed so the column density of amplifying gas is proportional to the number of overlapped clumps along the line of sight. This number is the result of a random process that yields a mean which is linearly related to the projected path length through the gas.

2. Description of the Model

2.1. Geometry

We propose that the conditions for the OH gas to produce maser amplification are satisfied within a thin ring structure with inner and outer radii R_i and R_o. Surrounding the OH, there is another concentric thin ring of radius R_c where most of the continuum emission is produced. In addition, free-free absorption cones along the axis are assumed to account for the east-west asymmetry in the brightness distribution. This situation is depicted in Figure 1 and the observed geometrical parameters are summarized in Table I.

The fact that the northern and southern compact masers seem to be elongated toward the east (Diamond et al., 1998; Trotter et al., 1989) is explained in our model

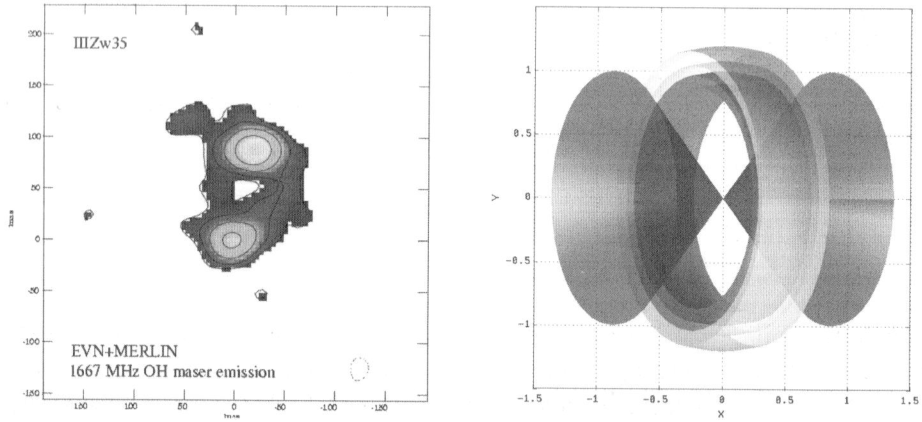

Figure 1. Left: Velocity averaged OH emission as seen at EVN + MERLIN resolution (Pihström et al., 2001). The contours are −1, 1, 2, 4 and 8 times the 3σ rms noise of 2.7 mJy/beam. *Right*: The proposed geometry. The OH gas is concentrated within the dark gray ring and most of the continuum is emitted within the surrounding light gray ring. The medium gray cones indicate the region where free-free absorption occurs. The units are normalized to the outer OH radius R_o.

TABLE I

Observed geometrical parameters

Parameter	Value	Reference
ΔR	3 Pc	Diamond et al., 1999; Trotter et al., 1997
ΔH	6 Pc	Diamond et al., 1999; Trotter et al., 1997
R_i	22 Pc	Trotter et al., 1997
R_o	25 Pc	Pihlström et al., 2001
R_c	50 Pc	Pihlström et al., 2001
θ	60 Deg	Pihlström et al., 2001

by the larger amount of continuum behind the OH in these regions. Accordingly, the apparent east-west extention of the feature corresponds to half of the ring width ΔH whereas the north-south extention matches the difference between the outer and inner radii ΔR. The angle θ between the axis of symmetry of the disk and the direction of the observer is estimated from the aspect ratio of the projected ring (Pihlström et al., 2001).

2.2. Clumpy Medium Analysis

The *clumpiness* of the medium is defined by the parameter $\bar{n}$ which defines the average number of clumps per cloud velocity width found in the radial direction within the ring. The average number of clumps in the direction of the observer is given by $\lambda(x, y) = \bar{n} \cdot L(x, y)/\Delta R$ where $L(x, y)$ is the projected path length through the OH gas. For any given line of sight, the number of observed spatial overlappings per velocity range is drawn from a Poisson random process of parameter λ with Probability Mass Function $P_\lambda(n)$. If all the clumps have the same optical depth τ_0, then the expected amplification G along the line is given by the sum:

$$G = \sum_{n=0}^{N} P_\lambda(n) \cdot e^{n\tau_0} \tag{1}$$

which in the limit of $N \to \infty$ converges to $e^{\lambda(e^{\tau_0}-1)}$ (See Conway et al. in this volume). This result is inappropriate in our context since the maximum number of aligned clumps N is upper bounded either by geometrical or statistical constrains. For these reasons, the mean does not constitute a good estimator for the observations. A better choice could be to use the median instead. A simplified approximation is to define an expected maximum N in order to truncate the sum in (1). Given a set of M independent realizations of the process, corresponding to M independent lines of sight within the beam, and requiring at least one occurrence of N overlaps to be observed, the expression for the maximum N becomes:

$$N_\lambda(M) = P_\lambda^{-1}\left(1 - \frac{1}{M}\right) \qquad (2)$$

where $P_\lambda^{-1}(x)$ is the Inverse Poisson Cumulative Distribution of parameter λ, and M is given by the ratio of the areas of the EVN + MERLIN to the VLBI synthesized beams when the clumps are assumed to be just resolved at the latter resolution. For the proposed geometry we define the following quantities and their observed values:

$$\begin{aligned} G_1 &= \sum_{n=0}^{N_1} P_{\lambda_1}(n) \cdot e^{n\tau_0} \approx 10 && \text{Gain in Region 1} \\ G_2 &= \sum_{n=0}^{N_2} P_{\lambda_2}(n) \cdot e^{n\tau_0} \approx 50 && \text{Gain in Region 2} \\ G_P &= e^{N_2\tau_0} \geq 500 && \text{Peak gain in Region 2} \end{aligned} \qquad (3)$$

where N_1 and N_2 are the maximum number of cloud overlaps in regions 1 and 2 given by (2) and $\lambda_2 = K \cdot \lambda_1$ with K being the ratio of the averaged path lengths within both regions. Each of these equations define a locus in the $\lambda_1 - \tau_0$ plane where the observed values are matched. The intersection of those loci corresponds to the

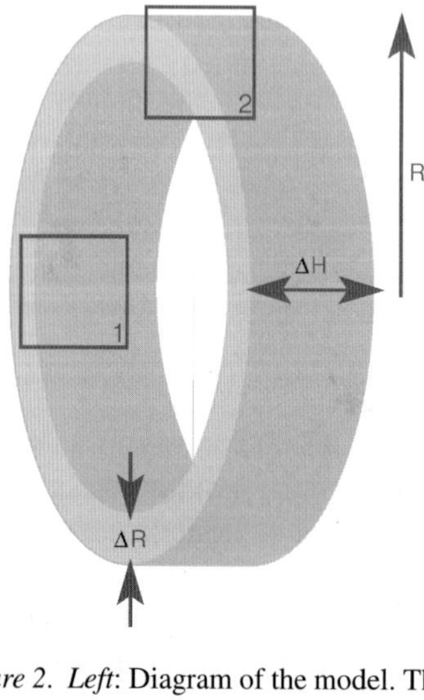

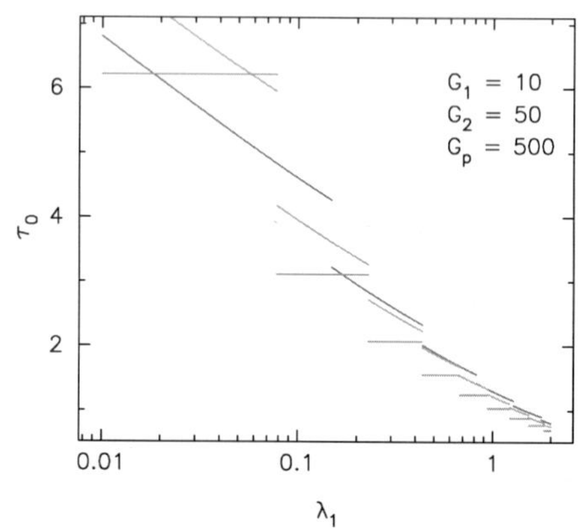

Figure 2. Left: Diagram of the model. The boxes indicate the positions of regions 1 and 2. The outer radius of the continuum ring is not visible in this drawing. *Right*: Solutions of equations (3) shown in the $\lambda_1 - \tau_0$ space. The intersection of the dark and light gray lines is the solution of the system.

possible combinations of λ_1 and τ_0 satisfying all three conditions simultaneously. Solving numerically, the solution is found to be $\lambda_1 = 0.67$ and $\tau_0 = 1.69$ which corresponds with a mean free path between clumps $\ell \approx 1.72$ pc.

3. Simulations

Using the parameters obtained in the previous section as input for Monte Carlo simulations, artificial spectra and brightness distributions were produced. Random generators were set up to reproduce the spatial distribution of the individual clumps and the pseudo-turbulent velocity field. The resulting XYZ cube was transformed into a XYV cube using the cascade model shown in Figure 3. The results are summarized in Figures 4 and 5.

Figure 3. Cascaded amplifiers model. The gain G_i of each amplifier is given by $\exp\{\tau_0 \cdot \exp((v - v_j)^2/2\sigma^2)\}$ Where $\sigma = 10$ km s^{-1} is the internal velocity dispersion of the clumps and v_j is the projected velocity of each clump constituted by a turbulent and a keplerian component. C_i are the continuum contributions between the clumps and A_{ff} is the free-free absorption attenuation. All the parameters are functions of (x, y).

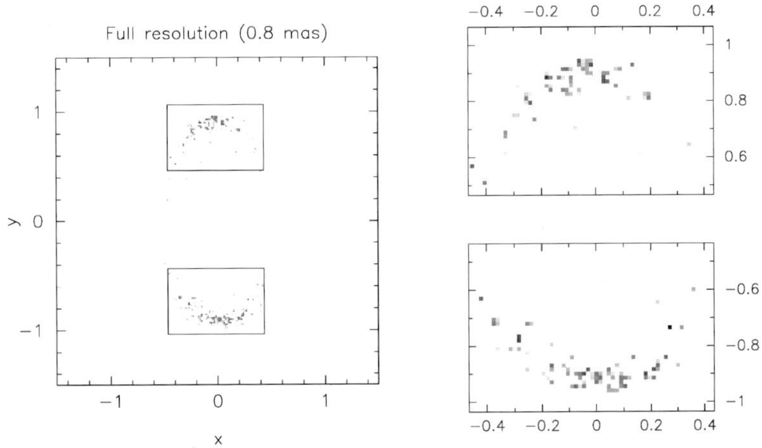

Figure 4. Integrated OH intensity $I(x, y)$. The XYV cube was averaged along all the spectral channels. The boxes indicate the northern and southern regions which are shown enlarged in the right panel. Both regions present strong compact emission dominated by a few very bright spots. In addition, the arc shapes observed with VLBI are fairly well reproduced. The gray scale has been normalized to the brightest point in the south.

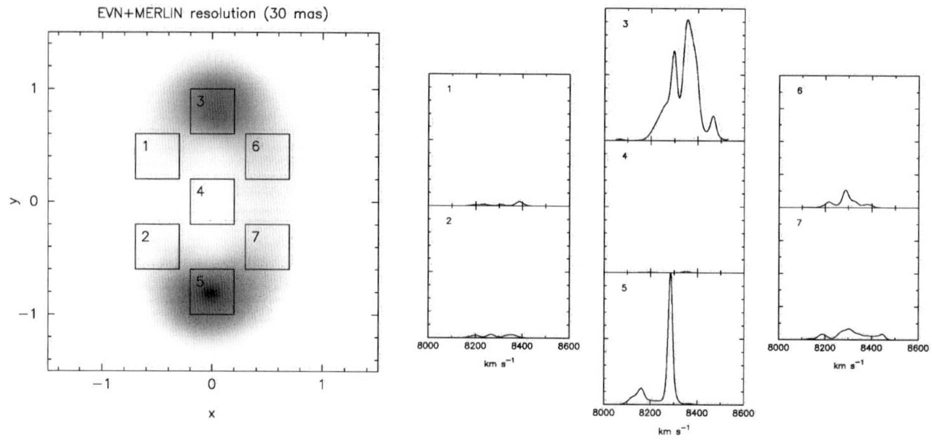

Figure 5. Left: Integrated OH intensity convolved down to EVN + MERLIN resolution. The *bridges* of smooth emission connecting the north and the south are clearly visible as well as the east-west asymmetry produced by the free-free absorbing cones. The grayscale has been normalized to the brightest point in the south. Boxes indicate the positions where spectra were taken. *Right*: The spatial location and relative widths of the spectral features correspond to those observed by Pihlström et al., 2001. The vertical scale has been normalized to the peak in the south.

4. Conclusions

The proposed geometry has managed to reproduce the observed brightness distribution and spectrum of the OH maser emission. The analytical approach in section 2 has been useful in fine tuning the Monte Carlo simulations. However, some aspects of the observed velocity field need to be studied. Work is still in progress connecting the obtained results with the physics of masers and the developing of a consistent formalism for the analysis of clumpy media.

References

Diamond, P.J., Lonsdale, C.J., Lonsdale, C.J. and Smith, H.E.: 1999, *ApJ* **511**, 178.
Klöckner, H.R., Baan, W.A. and Garrett, M.A.: 2003, *Nature* **421**, 821.
Martin, H.M., Sanders, D.B. and Hills, R.E.: 1984, *MNRAS* **208**, 35.
Montgomery, A.S. and Cohen, R.J.: 1992, *MNRAS* **254**, 23.
Rovilos, E., Diamond, P.J., Lonsdale, C.J., Lonsdale, C.J. and Smith, H.E.: 2003, *MNRAS* **342**, 373.
Pihlström, Y.M., Conway, J.E., Booth, R.S., Diamond, P.J. and Polatidis, A.G.: 2001, *A&A* **377**, 413.
Trotter, A.S., Moran, J.M., Greenhill, L.J., Zheng, X. and Gwinn, C.R.: 1997, *ApJ* **485**, L79.